D1724847

Hippokratis Kiaris

Understanding Carcinogenesis

Related Titles

Offit, K

Clinical Cancer Genetics: Risk Counseling and Management, 2nd Edition

2006
ISBN 0-471-45893-7

Debatin, K.-M., Fulda, S. (Eds.)

Apoptosis and Cancer Therapy

From Cutting-edge Science to Novel Therapeutic Concepts

2006
ISBN 3-527-31237-4

Harald zur Hausen

Infections Causing Human Cancer

2006
ISBN 3-527-31056-8

Liebler, D. C., Petricoin, E. F., Liotta, L. A. (Eds.)

Proteomics in Cancer Research

2005
ISBN 0-471-44476-6

Holland, E. C. (Ed.)

Mouse Models of Cancer

2004
ISBN 0-471-44460-X

Brenner, C., Duggan, D. J. (Eds.)

Oncogenomics

Molecular Approaches to Cancer

2004
ISBN 0-471-22592-4

Hippokratis Kiaris

Understanding Carcinogenesis

An Introduction to the Molecular Basis of Cancer

WILEY-VCH Verlag GmbH & Co. KGaA

The Author

Prof. Dr. Hippokratis Kiaris
Dept. of Biological Chemistry
Univ. of Athens Medical School
M. Asias 75
11527 Athens
Greece

Library of Congress Card No.: Applied for

British Library Cataloguing-in-Publication Data:
A catalogue record for this book is available from the British Library.

Bibliographic information published by Die Deutsche Bibliothek
Die Deutsche Bibliothek lists this publication in the Deutsche Nationalbibliografie; detailed bibliographic data is available in the Internet at <http://dnb.ddb.de>.

Printed in the Federal Republic of Germany.
Printed on acid-free paper.

Typesetting Hagedorn Kommunikation, Viernheim
Printing Betz Druck GmbH, Darmstadt
Bookbinding J. Schäffer GmbH, Grünstadt

ISBN-13: 978-3-527-31486-7
ISBN-10: 3-527-31486-5

Contents

Understanding Carcinogenesis. Hippokratis Kiaris

ISBN 3-527-31486-5

Part III Specific Topics

Part IV Unifying the Concepts

Part V Future Perspectives

To my beloved family, for their encouragement,
inspiration and support

To the anonymous post-doc who dedicated his or her best
and most productive years in cancer research

Understanding Carcinogenesis. Hippokratis Kiaris

ISBN 3-527-31486-5

Foreword

Recent progress in life sciences resulted in the accumulation of an enormous amount of information that because of its complexity is available only to specialized scientists with backgrounds specific to the corresponding fields. This is especially true for cancer biology that during the last decades evolved into a distinct entity, on the borderline of molecular biology and clinical oncology.

Dr. Hippokratis Kiaris undertook a difficult effort to write a book on molecular biology of cancer and I believe that he succeeded in it. His book covers the area between these two entities, the basic and the clinical. Without entangling the readers in technical details on pathways and specific genes and molecules involved in carcinogenesis that are familiar only to molecular biologists, or providing information regarding treatment modalities useful only to clinicians, he describes carcinogenesis in a simple and concise manner that is reflected by the title of the book: To understand and not only learn the details of the molecular basis of carcinogenesis. He builds up the book by introducing fundamental concepts related to the clonality of the tumors and the hypermutability of the cancer cells, continues with a description of genes involved in the process of carcinogenesis and concludes with more complex phenomena of tumor biology such as the role of the tumor stroma and the metastatic process. Specific topics such as pharmacogenomics and viral carcinogenesis are also well covered. In addition, he describes human and animal models of the disease, emphasizing their advantages and their limitations.

The book includes simple graphs that underline notions described in the text, a feature that is particularly useful to the reader. I strongly believe that this book will be a useful companion of the scientist who, having a background in basic molecular biology and genetics, desires to enter the field of molecular oncology, or of the clinician who wishes to understand the molecular basis of cancer development, I recommend it highly.

Andrew V. Schally
November 2005

Understanding Carcinogenesis. Hippokratis Kiaris

ISBN 3-527-31486-5

Preface

When attempting to write a new book, a technical one in particular, the author must make sure that such a book is needed and indeed it has to offer something in addition to that which is currently available in the literature. This is of particular importance in the field of cancer and tumor biology, in general, in which an enormous amount of information has been published. My stimulus for this effort was that despite the fact that information regarding cancer biology is published and is being published constantly, the material available falls into two major and distinct categories: it consists of either oversimplified books targeting the general audience or very detailed ones, at the level of reference books, intended to be read by specialized scientists such as oncologists and molecular biologists. There is an apparent gap In between. Someone with a basic knowledge of molecular biology, genetics and oncology has a very limited selection of books that are going to introduce him/her to the molecular biology of cancer without being consumed by details.

The timing of such a book is also important. Usually, cancer books state that currently we have started to understand the molecular basis of the disease. This is true, but yet somehow obsolete. Most likely the truth is that now we have begun to understand the actual complexity of cancer as the field enters the area of synthesis.

Throughout this book I assume that the reader has some knowledge of basic genetics as well as cell and molecular biology, while only the application of this knowledge to the cancer field is described. Whether I have succeeded will be judged by the reader.

The references and selected bibliographies are limited to some recent review articles written by world leaders in the field. Original research articles are also occasionally included on the basis of their historical significance and/or the absence of the appropriate description of the corresponding findings in reviews. I truly apologize to the scientists not mentioned.

Understanding Carcinogenesis. Hippokratis Kiaris

ISBN 3-527-31486-5

I wanted to thank the postgraduate students for their stimulating discussions that shaped the structure of this project and prompted me to undertake this effort, and the editorial staff at Wiley-VCH who were invaluable in producing the final book.

Athens
September 2005

Part I
Basic Aspects

Understanding Carcinogenesis. Hippokratis Kiaris

ISBN 3-527-31486-5

1 Some Introductory Concepts in Tumor Biology: Clonal Evolution and Autonomy versus Non-autonomy of Cancer Cells

Cancer has a Genetic Basis

Cancer is a genetic disease. This statement is a consensus that can be found in many cancer books and science articles, and reflects the fact that changes at the level of the genetic material, the DNA, are responsible for disease development. In addition to the hereditary types of the disease in which this genetic basis is apparent and self-evident, all types of sporadic cancer progress through alterations at the level of the cell's genome. Tumor-causing viruses, mutagens and all other factors associated with carcinogenesis either directly or indirectly affect the DNA at a certain time.

Clonality

Cancer is also clonal. Indeed, cancer is initiated by a single cell that carries (inherits or is being targeted by) a mutation. Subsequently, this cell expands clonally. However, this expansion is continually subjected to selection as the cancer cells constantly accumulate mutations that change their behavior. The mutations that contribute the more aggressive characteristics, like rapid growth and resistance to certain therapeutic modalities, are those that predominate. Therefore, a tumor becomes clinically worse with time. By examining the genotype of the cancer cells in a given tumor it is possible to identify such mutations, and to trace their physical history based on the subset and the histopathology of the cells that carry a specific genetic lesion. This "dynamic" nature of cancer cells is responsible for the failure of conventional therapies and our inability to develop appropriate diagnostic tools that can be applied to even a single distinct type of cancer.

Autonomy of Cancer Cells

Cancer cells are also characterized by increased autonomy compared to normal cells of the same origin. Changes in the environment of the cancer cells generally make them more tolerant than their normal counterparts. This property of malig-

Understanding Carcinogenesis. Hippokratis Kiaris

ISBN 3-527-31486-5

nant cells is reflected in the fact that under culture conditions it is much easier to grow cancer cells than normal cells, while culturing the latter always bears the danger of selecting for the cells that become transformed. However, despite this apparent property of cancer cells, continuous cancer growth exhibits a high dependency on the microenvironment to a degree that current opinion views malignant (and not only) tumors as an heterologous entity of different (genetically and histologically) cell types that interact reciprocally with each other.

Heterogeneity of Cancers

Finally, cancer is not one, but many diseases. Indeed, countless combinations of genetic lesions contribute to the development of virtually unlimited clinical entities that, while they are all referred to as malignant tumors, exhibit dramatic differences in their histopathological and clinical characteristics. Even tumors of exactly the same histopathology and classification may exhibit quite important differences in their clinical outcome.

Although these statements are in principle accurate, they carry certain limitations with important consequences in tumor biology, as will become apparent in the following chapters. We will try to view carcinogenesis in a simplistic manner that, however, permits the perception of its complexity and, were possible, its unity.

2
The Cell's Life and Death: Cell Cycle, Senescence and Apoptosis

One of the most characteristic properties of cells that constitute a malignant tumor is the deregulation in the equilibrium between cell division versus cell death. During physiological conditions this equilibrium is also frequently subjected to imbalances towards either the side of the division or towards the side of cell death in a manner that is regulated very precisely. For example, the growth of tissues is achieved by inducing more cells to enter into an active proliferating state as compared to the fraction of cells that do not proliferate. On the contrary, cell death is favored under other conditions, as happens, for example, during the stage of involution of the mammary epithelium when mammary epithelial cells are subjected to strong signals that induce their death. In this case, up to 90% of the (mammary epithelial) cell population dies following lactation, resulting in the re-organization of the mammary gland and simulating a virgin-like state upon the completion of this process.

In cancer, however, these processes are deregulated and tumors eventually develop. Cancer cells use the signaling pathways that control cell division and death in their favor, and grow at the expense of the surrounding normal cells. At the cellular level these processes correspond to specific stages in the individual cell's life termed the "cell cycle", i.e. the procedure that upon its completion the cell divides and is actively induced when the cells become cancerous. Another stage during the cell's life is that of "senescence", which reflects the state of cellular aging. Senescence can be induced by exogenous signals or endogenously after extensive replication beyond the cell's capacity ("replicative senescence"). This stage reflects the state of metabolically active, but nondividing cells. Exit from senescence is usually followed by the death of the cell. The death of the cell death is a very precisely controlled process and can be "apoptotic" or "necrotic".

As might be expected, these stages are tightly regulated and are characterized by a high interdependence. For example, arrest in the cell cycle may induce senescence that can eventually end up as apoptosis. Alternatively, progression of the cell cycle requires suppression of the apoptotic pathways. In carcinogenesis, continuous cycling of the cells is needed, which in turn requires that the cells bypass the *crisis* that follows after escaping from prolonged senescence, as well as the inhibition of the pathways that induce apoptotic cell death.

Understanding Carcinogenesis. Hippokratis Kiaris

ISBN 3-527-31486-5

The importance of these pathways in carcinogenesis is two-fold. It is important to elucidate the underlying mechanisms (i) at the basic level that aims at the understanding the mechanisms of the disease and (ii) at the level of clinical practice considering that appropriate interventions in each of these states/processes may represent promising therapeutic targets or have other prognostic significance. For example, induction of apoptosis and inhibition of the cell cycle may represent desired targets of therapy, while induction of senescence (instead of apoptosis) during therapy may be an important reason for the failure of certain therapeutic modalities. Some descriptions of these basic processes and of their molecular regulation follow.

Cell Cycle

It is not difficult to imagine why cell cycle regulation plays a critical role in cancer development. Considering that cancer virtually represents the result of the uncontrolled excess of cell proliferation over cell death, it is easy to understand why dysregulation of cell cycle checkpoints is essential for neoplastic growth. While a detailed description of the cell cycle at the molecular level is beyond the scope of this book, it is important to note some basic aspects of how cell cycle progression is regulated.

The cell cycle is shown schematically in Fig. 1. It is divided into four major phases termed sequentially M, G_1, S and G_2. According to a simplified view, mitosis and cytokinesis occur during the M phase, while DNA replication is performed during the S phase. G_1 and G_2 represent preparatory phases in which the molecular events that prepare the cell for entry into the next phase occur. These phases, especially G_1, represent the stages that under physiological conditions are subjected to regulation under the response of the factors that induce or suppress cell cycle progression. An essential regulatory role is played by proteins called cycle-dependent kinases (CDKs). Three of them are subjected to sequential activation and trigger transition into subsequent phases of the cell cycle. CDK4/6 governs exit from G_0 (arrested cells) and is activated upon exposure to growth stimuli. CDK1 is responsible for the exit from G_1 and entrance into S, while CDK2 is responsible for the G_2/M transition. Molecular targets of these CDKs as well as regulators of their activity have been identified as oncogenes and tumor-suppressor genes (TSGs). For example, cell cycle progression is induced by the binding of the CDKs to specific proteins called cyclins. Many of these cyclins have been identified as potential oncogenes, as they are capable of inducing tumor growth in experimental animals and they are also frequently upregulated in various primary human cancers. On the contrary, proteins that inactivate CDKs by binding to them, resulting into the inhibition of the formation of complexes with specific cyclins, have been identified as TSGs. Two major classes of such CDK inhibitors (CDKIs) have been identified, including the proteins p15, p16, p18 and p19 (INK family), and p21 and p27 (CIP/KIP family). It has to be noted that the function of the CIP/KIP family members appears more complex

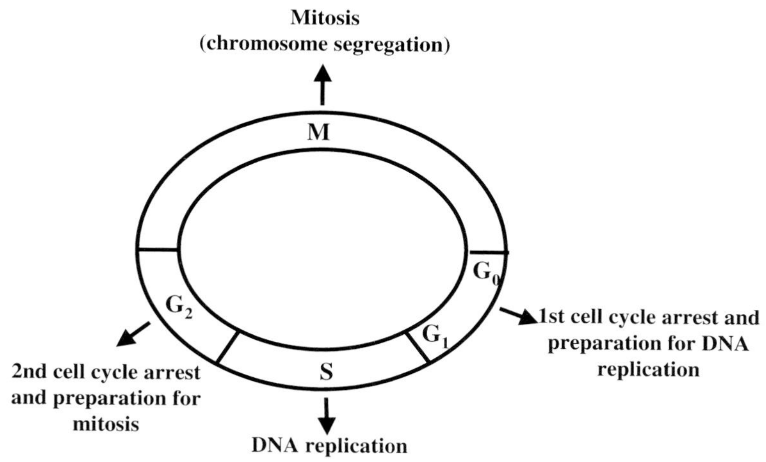

Fig. 1 The major stages of the cell cycle. G_0/G_1 may arbitrarily be considered as the first stage of the cell cycle. Quiescent cells are usually "stopped" in this phase and upon changes in their microenvironment (growth factors, hormones, etc.) proceed to the next stage or, alternatively, they may commit suicide. In this stage the cells are prepared for the next stage during which the DNA replicates (S). Successful DNA replication is followed by a second preparatory/arrest stage (G_2) that introduces the cell into mitosis (M) – the stage at which DNA takes the form of chromosomes. Subsequently, the mitotic spindle is created and the duplicated chromosomes divide into daughter cells. The end of mitosis marks the completion of a round of cell division. While under physiological conditions only a few cells are actively dividing (with certain exceptions) and, thus, the M phase occupies only a small fraction of the cell's life, in cancer cells a considerable increase in the fraction of actively dividing cells occurs, which is reflected in the frequency at which mitotic figures can be seen.

as their role depends on their levels – it can be either negative at high levels or positive at lower levels. This association between the precise level of expression of these proteins and the "functional output" is also reflected by the finding that they may operate as haploinsufficient genes in carcinogenesis (see Chapter 4).

Two other TSGs with an important role in cell cycle regulation are the nuclear proteins p53 and Rb. Among its many functions (that notably are not limited to the regulation of cell cycle progression), p53 activation induces the CIP/KIP family member CDKI p21, thus inhibiting G_1/S transition, while activation of Rb suppresses the activity of transcription factors that regulate the expression of specific cyclins and other regulatory proteins. Importantly, inactivation of p53, Rb and INK family members (especially p16) represents a common feature of various primary human tumors. Thus, the signaling network(s) that actively suppress cell cycle progression must be inactivated in carcinogenesis.

Senescence

Under physiological conditions the cells replicate as dictated by their endogenous differentiation program and the availability of growth factors in their microenvironment. However, some exogenous signals such as the presence of specific inhibitors or the absence of positive (proliferative) signals may inhibit the progression of the cell cycle and induce cell cycle arrest. As soon as these signals are no longer present, the cell continues to proliferate. In certain instances, however, a cell may undergo irreversible growth arrest – a state that has been termed senescence. In this case, even supplementation of growth factors into the cell's microenvironment is insufficient to promote the re-entry into the cell cycle. Factors that may induce senescence are the recognition of DNA damage or the shortening of the telomeres (replicative senescence). Occasionally, senescence may also be induced by exogenous factors that rapidly promote G_1 arrest, entering a state that simulates senescence in terms of morphology, molecular profile as well as in the inability to undergo proliferation even after re-exposure to mitogens. This state has been termed "STASIS" (STress or Aberrant Signaling Induced Senescence).

Senescence is quite important for the carcinogenic process. Senescent cells may produce secreted factors that stimulate the growth of adjacent cells. This is of particular importance in tumor stroma considering that senescent fibroblasts create a pro-oncogenic environment that promotes cancer growth. A similar mechanism is also responsible for some cases of anticancer therapy failure. Anticancer agents may induce senescence in some cells, either normal or malignant, that by then secreting growth factors stimulate the growth of the cancer cells that were supposed to die as a result of the treatment. It appears that in the context of viewing the tumor as an entity, senescence reflects the response of a fraction of the cells that sacrifice their proliferative capacity in favor of other cells that by receiving the signals elicited from them increase their probability of survival.

Studying the mechanisms of senescence in fibroblasts has proven quite informative. In this cell type, and probably in others as well, it has been shown that senescence occurs in two stages designated as the mortality stages 1 and 2 (M1 and M2). Replicative senescence corresponds to M1. Cells that bypass M1 divide until M2, the stage that corresponds to a crisis at which cells die by mitotic catastrophe (the cumulative result of many abnormal mitoses), which in turn may trigger apoptotic and nonapoptotic cell death. Benign lesions appear to bear a higher fraction of senescent cells than corresponding malignant lesions. Therefore, bypassing senescence represents an important mechanism for malignant conversion.

At the mechanistic level, several genes have been implicated in the induction and maintenance of senescence. Among them we briefly mention the p53 tumor suppressor and its downstream target CDKI p21 that are activated, usually transiently, following the signal that triggers senescence. Subsequently, and when the activity of these genes decreases, another CDKI, p16, becomes constitutively upregulated, suggesting its importance in maintaining growth arrest in senescent cells (Fig. 2). Importantly, all these three genes are tumor suppressors, exemplifying the significance of senescence as an anticarcinogenic mechanism.

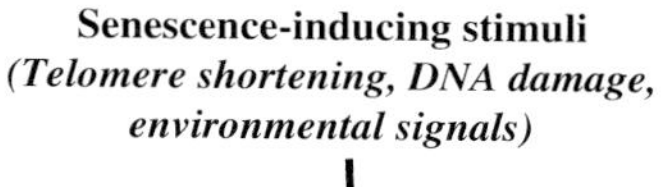

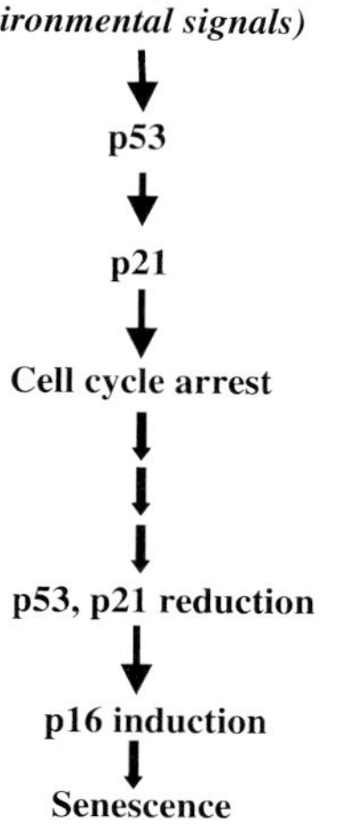

Fig. 2 Simplified view regarding a predominant pathway inducing senescence. The cell receives the signal that induces senescence and initially responds by activating p53, which in turn induces p21 expression. This results in cell cycle arrest. After prolonged arrest, p53 and p21 decrease, p16 is upregulated, and the cell enters senescence.

Apoptosis

When the state of the cell does not warrant its further survival (or when certain physiological conditions dictate so), the cell must die by committing suicide. This suicidal process is very tightly regulated and is termed "programmed cell death" or apoptosis.

It is generally accepted that programmed cell death is induced by two major pathways – an extrinsic and an intrinsic pathway. The extrinsic pathway is initiated by specific death receptors that are activated by their ligands which are members of the tumor necrosis factor (TNF) family of ligands. These ligands include TRAIL (TNF-related apoptosis-inducing ligand) – a promising target for the development of novel anticancer strategies. The intrinsic pathway involves the mitochondrion. In both cases the execution of the apoptotic response requires the activation of specific proteases, termed caspases (Fig. 3). The importance of mitochondrion in the execution of apoptosis is in contrast to the conventional notion that viewed this organelle only as the energy factory of the cell that had as its single role the generation of energy in the form of ATP. Recent data have demonstrated that upon receipt of specific signals, mitochondria release proapoptotic factors such as the cytochrome *c*. An important regulatory role in these processes is played by the *myc* oncogene (see Chapter 3) and the Bcl-2 family member Bid. Upon release into the cytoplasm, cytochrome *c* forms complexes with apoptosis activating factor-1 (Apaf-1), ATP and procaspase-9. The formation of this complex, also termed "apoptosome", results in the activation of procaspase-9 to caspase-9 (an initiator caspase). Subsequently, other caspases, such as caspase-3, -6 and -7 (executioner caspases) are activated, and the proteolytic cascade is initiated, resulting in the acquisition of the characteristic morphology of the apoptotic cells mani-

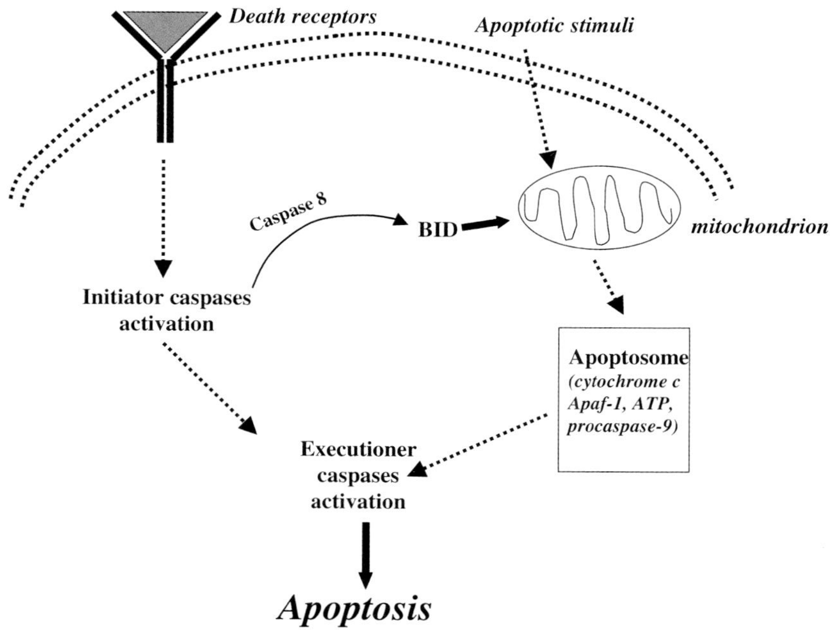

Fig. 3 Two major apoptotic pathways have been identified – extrinsic and intrinsic. The first is largely controlled by the activation of specific membrane receptors designated as death receptors. For the second, the mitochondrion plays a central role by eliciting certain proapoptotic factors. While usually described as independent entities, crosstalk between these pathways has been described. For example, caspase-8 (an initiator caspase involved in the extrinsic pathway) also cleaves BID, which in turn activates the mitochondrial pathway.

fested by chromatin condensation, nuclear shrinkage, DNA fragmentation and others. Finally, the remains of the dying cell are engulfed by phagocytes. Other apoptotic pathways that are caspase independent have also been proposed.

The majority of the genes that control the apoptotic response are TSGs (see Chapter 4) and proteases, such as the caspases that execute the apoptotic program (see above). Of special emphasis in the regulation of apoptosis, especially the form that is tightly related to the mitochondrion, is the Bcl-2 family members that are thought to supervise the apoptotic response. These genes cannot be defined as oncogenes or TSGs in the formal manner since they do not display transforming or tumor-suppressive activities in the corresponding standard assays. On the contrary, Bcl-2 family members are proapoptotic (Bcl-2, Bcl-X_L and Mcl-1) or antiapoptotic (Bak and Bax). They contain the BH (Bcl-2 homology) domains (BH1–4), with BH4 being the domain that is present only in the antiapoptotic members of the family and BH3 being the domain that is present in all Bcl-2 family members. The balance between the pro- and the antiapoptotic members of this family plays an important role in the regulation of the execution of the apoptotic program.

Malignant conversion and the acquisition of the tumorigenic phenotype require the adoption of resistance to apoptosis. This becomes apparent from the fact that overproliferation alone is insufficient for cancer formation because mitogenic signals elicited by various oncoproteins almost always induce apoptosis directly or indirectly. Thus, survival signals must be elicited at the same time as the overproliferating ones. These survival signals are produced by the various growth factors that are frequently overexpressed in cancer cells and result in the activation of specific signaling cascades, by a series of phosphorylations. An example is offered by the signaling cascade related to the activation of phosphatidylinositol-3-kinase (PI3K) (see ErbB oncogenes in Chapter 3). Importantly, the prediction that certain phosphatases may have tumor-suppressor activity by inhibiting these survival signals was confirmed following the identification of the PTEN (phosphatase and tensin homolog deleted on chromosome 10) tumor suppressor that encodes for a phosphatase that antagonizes the action of PI3K.

"Anoikis" represents a special form of apoptosis that is related, and indeed triggered, following the formation of inappropriate contacts between cells and the matrix. Anoikis is thought to play an important role in the metastatic process.

Apoptosis is also induced by the shortening of telomeres (telomere erosion). The p53 tumor suppressor, which is involved in the regulation of the apoptotic response elicited especially by various types of genotoxic stress (Fig. 4), plays an important role in this form of apoptosis. However, bypassing the p53 dependency to apoptosis, due to telomere erosion eventually also causes p53-independent apoptosis, most likely due to mitotic catastrophe.

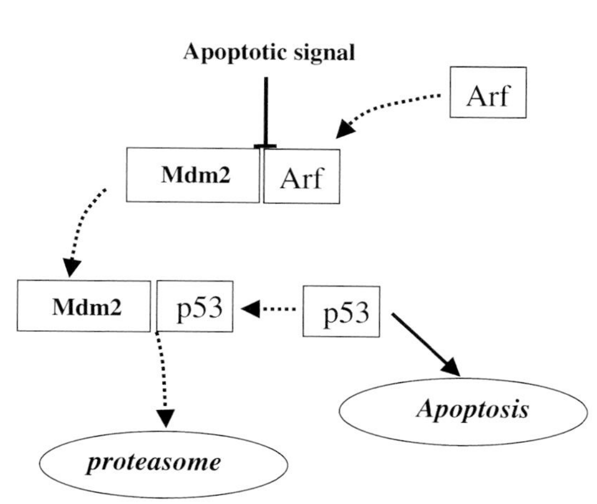

Fig. 4 The role of p53 in regulation of the apoptotic response. The activity of p53 is controlled by a network of positive and negative regulators such as MDM2 and Arf. In the absence of apoptotic signals, under physiological conditions, p53 binds to MDM2 – a ubiquitin ligase that inactivates p53 by bringing it to the proteosome resulting in its proteolytic cleavage. However, when apoptotic signals are present, Arf is induced and binds to MDM2, dissociating it from p53. In this case, p53 is free and is able to induce the apoptotic cascade by stimulating the expression of downstream target genes.

Cell Death: Apoptosis versus Necrosis

Although several subcategories with defined morphological characteristics have been suggested, two major categories of cell death can be defined – "necrotic" and "apoptotic". The major difference is that in the necrosis the cell's morphology is random and it is a passive process that most likely does not require caspase activation, contrary to apoptosis in which the cells (as already mentioned) acquire a nonrandom morphology. Furthermore, apoptosis requires energy consumption and proceeds through the regulated activation of caspases. Recently, however, the causative role of caspases in programmed cell death has been debated as it was suggested that their role is at the level of defining rather than inducing the precise characteristics of cell death in cells that have already been committed to die. Furthermore, necrotic cell death occurs more frequently under pathological conditions, whereas apoptosis occurs under physiological conditions and at certain developmental stages, but may be also detected under pathological conditions as well.

The molecular regulation of apoptosis is described in some detail elsewhere (see Chapters 4 and 13). However, an important aspect of this process is viewing this decision as a switch between necrosis and apoptosis. Experimental evidence showed that the same cells can undergo either apoptosis or necrosis under exposure to the same stimulus, by manipulating some parameters of the cellular physiology. For example, by reducing the levels of intracellular ATP it was shown that certain cells may undergo necrosis and not apoptosis, as happens when ATP levels are higher. This finding clearly illustrates that apoptosis is a process that demands energy.

In the chapters that follow it will become clear that throughout the progression of carcinogenesis, from the very early stages where a tumor is initiated as a benign lesion until the late stages when a tumor becomes metastatic, a constant interplay occurs between cell cycle progression, entrance into senescence and initiation of apoptosis. It is important to view this interplay as a continuous decision process, always keeping in mind that it is the same genes that regulate each of these processes and increasing the complexity of these molecular decisions. Modulating the output of these decisions promises important therapeutic interventions towards "intelligent" anticancer therapy.

Bibliography

Jäättelä M. Multiple cell death pathways as regulators of tumour initiation and progression. *Oncogene* **2004**, *23*, 2746–2756.

Nicotera P, Melino G. Regulation of the apoptosis–necrosis switch. *Oncogene* **2004**, *23*, 2757–2765.

Shay JW, Roninson IB. Hallmarks of senescence in carcinogenesis and cancer therapy. *Oncogene* **2004**, *23*, 2919–2933.

Tachibana KK, Gonzalez MA, Coleman N. Cell-cycle-dependent regulation of DNA replication and its relevance to cancer pathology. *J Pathol* **2005**, *205*, 123–129.

Porath I-B, Weinberg RA. The signals and pathways activating cellular senescence. *Int J Biochem Cell Biol* **2005**, *37*, 961–976

Igney FI, Krammer PH. Death and anti-death: tumor resistance to apoptosis. *Nat Rev Cancer* **2002**, *2*, 277–288.

Part II
Genes Involved in Carcinogenesis

Understanding Carcinogenesis. Hippokratis Kiaris

ISBN 3-527-31486-5

3
Oncogenes

The study of oncogenes, along with that of tumor-suppressor genes (TSGs), occupies a special place in cancer research because it is historically responsible for giving the field its molecular/mechanistic dimension. While the operation of tumor-suppressor mechanisms, reflecting the existence of TSGs, has been shown by genetic studies (i.e. retinoblastoma and Knudson's two-hit hypothesis, see Chapter 4), early experiments with oncogenes demonstrated the causative relationship between specific genetic alterations and carcinogenesis.

Which genes are oncogenes? In order to be classified as such, they must be able to cause the malignant transformation of noncancer cells. There are two important, common and essentially diagnostic assays performed in cells, usually immortalized fibroblasts, in order to test if they have been subjected to malignant transformation after (over)expression of the gene that is under investigation:

(1) Anchorage independence and contact inhibition. Noncancer cells, even including those that are immortalized, require attachment to a basement membrane in order to grow in culture and certainly do not grow in semisolid media. Continuing culture under proper conditions results in the inhibition of their proliferation as soon as they reach confluency and each cell is in contact with another. However, cancer cells are able to form foci, as they can grow on top of each other (Fig. 1). Furthermore, when the cells are forced to grow in media that contain agar (or agarose) at a low concentration, usually below 1% under conditions where the environment of the cells consists of a soft gel, only transformed cells are able to grow, whereas noncancer cells do not. The term "noncancer" instead of normal is used to exclude the immortalized cells.

(2) The second assay involves the evaluation of their ability to grow in nude (or any other strain of immunoincompetent) animals, usually subcutaneously. Therefore, following the specific genetic manipulation that aims to overexpress the gene to be tested in noncancer cells (again, usually immortalized fibroblasts) the cells are injected into the experi-

Understanding Carcinogenesis. Hippokratis Kiaris

ISBN 3-527-31486-5

Fig. 1 Contact inhibition and foci formation. Contrary to normal and immortalized cells (left), cancer cells in culture (right) do not stop proliferating when they reach confluency and they generate foci as they grow on top of each other. This property is associated with the metastatic and invasive ability they display *in vivo*.

mental animals and the injection sites are observed for about 1–2 months for the onset of tumors.

How accurate and precise can these tests be? The answer is that they are relatively good and predictive; however, their limitations should always be kept in mind. For example, as will become apparent at the following chapters, one important parameter is the cellular and tissue context. A given gene may be potently oncogenic for one tissue or cell type and moderately oncogenic (and in certain cases even tumor suppressive) for another. Furthermore, the tissue context plays an important role as tumorigenicity, exemplified by the ability of the cells to grow in mice, is not a general property of all cancer cells. For example, established cell lines from some mammary carcinomas, e.g. MCF7 cells, can only grow in nude mice if hormone supplementation or coinjection of stromal fibroblasts along with the cancer cells is performed. Another example is given by the ability of embryonic cells to give rise either to cancers (teratocarcinomas) or normally developing embryos, depending on the site of injection.

Even after taking into consideration these limitations, some conceptual problems still persist with the application of these assays, related to how accurately they simulate the real conditions in human cancer. First of all the forced expression of the gene to be tested achieves a degree of overexpression that surpasses (occasionally by orders of magnitude) the endogenous expression of the candidate oncogene. Furthermore, by definition, the oncogenic potential should be exerted on normal cells and not immortalized cells. This is probably the reason why the successful performance of these assays, i.e. causing cancer by the genetic manipulation of a single locus, does not contradict the widely accepted notion of the multistage nature of carcinogenesis. Thus, immortalized cells have already acquired some mutations that make them prone to malignant transformation. Furthermore, it is noteworthy that rodent cells are generally more prone to carcinogenesis than human cells, at least under the conditions of the aforementioned assays.

Recent progress in transgenic technology has partially solved some of these obstacles by providing more realistic models to study carcinogenesis, in a context that mimics human disease. It should be mentioned that the complexity of carcinogenesis renders the performance of a single assay to diagnose the oncogenic potential of a gene inappropriate. Rather, it is a combination of assays that may characterize a gene as being an oncogene. Finally, it has to be kept constantly in mind that the classification of genes into specific and well-defined categories constitutes an oversimplification that is probably didactic and appropriate for introductory purposes, but in reality is illusive.

Contrary to the TSGs that classically, with only a few exceptions, encode for nuclear proteins that are generally regulators of the cell cycle, oncogenes can be nuclear (e.g. various transcription factors), cytoplasmic (e.g. soluble enzymes), membrane proteins (e.g. receptors) or secreted proteins that operate as growth factors. Genetically, they are frequently subjected to mutations that result in the generation of novel superactive alleles that encode for proteins with a more potent action than that of the endogenous, physiological allele. This can be seen by point mutations that generate constitutively active forms of the protein. Classical examples are the codon 12, 13 and 61 point mutations in the members of the *ras* family of oncogenes. In certain cases, such mutations are detected very frequently in specific neoplasms, such as the pancreatic cancers that bear K-*ras* mutations in almost 90% of cases. Alternatively, they do not undergo activation through mutations, but by simple overexpression of the wild-type allele that results in quantitative instead of qualitative changes in the oncogene's expression. This overexpression can be the result of transcriptional activation or amplification. Examples are the members of the *ras* family of oncogenes and the *myc* oncogene, respectively. Of course, a combination of more than one activating mechanisms can frequently be operating. For example, a mutant *ras* allele can also be amplified, while amplification in the *myc* oncogene can also be accompanied be transcriptional activation.

An important aspect of the biology of oncogenes is the fact that many of them have been (originally in many cases) identified as the transforming proteins of certain cancer-causing viruses. Therefore, these cellular homologs of the viral oncogenes (v-*onc*) have been designated as c-*onc*. Other highly oncogenic viral proteins with no considerable homology to cellular genes are the T antigens of various oncogenic viruses. Those proteins that act by binding and inactivating endogenous TSGs will be discussed in greater detail in Chapter 16.

Below, we will mention some examples of oncogenes that provide examples of the various mechanisms by which an oncogene can acquire its oncogenic potential during carcinogenesis. These oncogenes are *myc*, the ErbB2 family of tyrosine kinases and the members of the *ras* family. Finally, we will also mention a novel category of transcripts – some short RNAs (microRNAs) that exert oncogenic potential without encoding for specific proteins.

myc Oncogene

myc is among the most, if not *the* most, potent human oncogene in terms of its ability to elicit tumorigenesis in a variety of *in vitro* and *in vivo* assays and cellular contexts. The original implication of *myc* in carcinogenesis was suggested following its identification as a target of translocation in primary Burkitt's lymphomas that results in its overexpression. Subsequently, it became clear that overexpression of *myc* is common in human tumors, by mechanisms that involve its transcriptional activation, without being accompanied by amplification. Virtually all signaling pathways activated in carcinogenesis result in the direct induction of *myc* expression. Those pathways include receptors for tyrosine kinase growth factors, β-catenin and others.

c-*myc* is the prominent member of a family of genes that also includes L-*myc*, and N-*myc*, and which encode for nuclear proteins with the characteristic structure of basic helix–loop–helix leucine zipper (bHLH-Zip) transcription factors. Myc protein(s), upon hetero-dimerization with another small protein named Max, is able to bind specific sequences in the DNA and operate as a transcription factor, while antagonizing activity to Myc is exerted by the Mad–Max and Mnt–Max complexes that actively repress gene transcription (Fig. 2). The consensus sequence recognized by the *myc*-containing transcription complex contains the CAYGTG sequence. The short length of this sequence predicts its abundant presence in regulatory regions. Indeed, functional CACGAG sequences are detected frequently in promoter regions of many different genes, yet the specific targets that are responsible for the potent oncogenic activity of *myc* remain unknown. It is notable that about 15% of all human genes bind *myc*. The complexity of *myc* function is further increased by experimental evidence suggesting that *myc*, in addition to its action in regulating gene transcription, is also able to elicit divergent responses by alternative mechanism(s) that do not involve its action as a transcription factor. *Myc* was also identified as a direct positive regulator of hTERT, the gene encoding for the catalytic subunit of telomerase, implicating *myc* in the regulation of immortalization and providing additional clues regarding the pleiotropy of its action (i.e. the fact a single mutant gene has the ability to cause multiple mutant phenotypes). It has also been suggested that the potency of *myc*-elicited oncogenic signals is largely due to their ability to cause genomic instability, possibly through the induction of reactive oxygen species.

At a mechanistic level it has been shown that *myc*'s consequences in a cell show an intriguing paradox. Apart from its undoubted function as a pro-oncogenic factor, which is exemplified by its ability to cause stimulation of angiogenesis, induc-

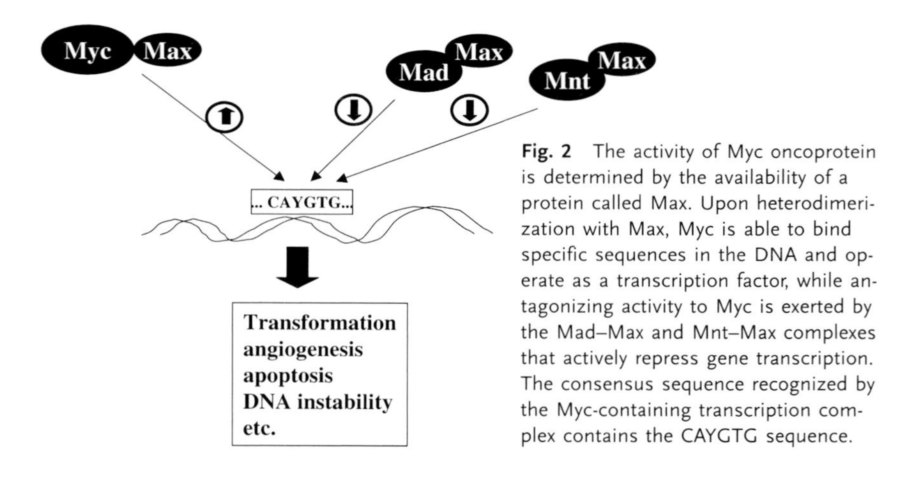

Fig. 2 The activity of Myc oncoprotein is determined by the availability of a protein called Max. Upon heterodimerization with Max, Myc is able to bind specific sequences in the DNA and operate as a transcription factor, while antagonizing activity to Myc is exerted by the Mad–Max and Mnt–Max complexes that actively repress gene transcription. The consensus sequence recognized by the Myc-containing transcription complex contains the CAYGTG sequence.

tion of genomic instability, stimulation of cell cycle progression and malignant transformation, *myc* is also capable of causing apoptosis. This, which was shown to be an intrinsic property of other oncogenes as well, suggests that in order to cause cancer, the simultaneously induced *myc* proapoptotic pathways must somehow be suppressed.

This proapoptotic function of *myc* should be viewed in two different ways. It can be a proper physiological function that has implications during normal development or, alternatively, it represents a physiological and protective response of the cell against the stress elicited by the high *myc* levels that render a cell susceptible to malignant transformation. It is likely that the identification of the precise molecular targets of *myc* and its integration into oncogenic pathways will shed light on this question.

For example, it has been shown that *myc* suppresses the p53-induced activation of $p21^{Waf1}$, the CDK inhibitor that is largely responsible for the cell cycle arrest effects of p53. Thus, its suppression mediated by *myc* favors the alternative response elicited by p53, which is proapoptotic. Additional examples on how *myc* regulates apoptosis are provided by its interplay with the Bcl-2 family of genes that regulate cellular suicide and the release of proapoptotic factors such as the cytochrome *c* from the mitochondrion.

The complex role of *myc* in oncogenesis, and its causative implication in such diverse pathways as malignant transformation, induction of apoptosis and stimulation of cell proliferation exemplifies the complexity of the oncogenic process.

ras Oncogenes

ras oncogenes are among the most prominent oncogenes and are widely used as transforming proteins. They control divergent processes such as proliferation, angiogenesis and malignant transformation, operating as signaling "hubs" between different signaling pathways. Upon their activation they initiate a multitude of signaling cascades, resulting in the regulation of growth, development and, importantly, oncogenesis. The oncogenic activity of the *ras* family genes has been clearly demonstrated in a variety of experimental systems, both *in vitro* causing malignant transformation and *in vivo* in mutant (transgenic) mice which developed cancers in the tissues to which activated *ras* proteins were targeted.

The members of the *ras* family are H-*ras*, K-*ras* and N-*ras*. They encode for small GTPases and are the founding members of a superfamily of GTPases containing more than 150 members in humans, while about 35 of them are closer to the *ras* family structurally (up to 55%) and functionally. The product of the *ras* family genes is a 21-kDa protein, $p21^{ras}$ that is located at the inner surface of the plasma membrane and operates as a switch between GTP/GDP. GTP-bound Ras (Ras-GTP) is the active form of the protein, whereas GDP-bound Ras (Ras-GDP) represents the inactive form. Upon reception of the signal that will activate Ras, specific guanine nucleotide exchange factors (GEFs) induce the formation of Ras-GTP (active), while at the same time GTPase-activating proteins (GAPs)

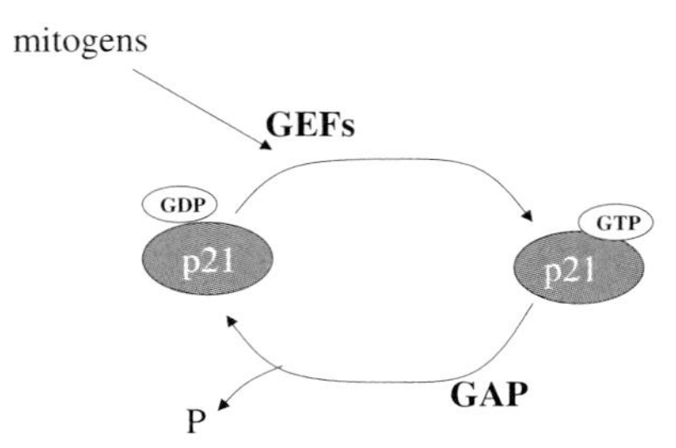

Fig. 3 Activation and inactivation of Ras proteins. Mitogens induce the exchange of GDP from GTP in Ras proteins by a mechanism that involves the action of specific GEFs. Subsequently, GTP bound onto the Ras proteins is hydrolyzed into GDP and P, and Ras proteins become inactive. This hydrolytic step involves the action of specific GTPase-activating proteins that stimulate the GTPase activity of Ras.

stimulate the GTPase activity of Ras, inducing the transition of Ras-GTP to Ras-GDP and thus inactivating the complex (Fig. 3).

Importantly, mutant forms of the *ras* family genes are detected relatively frequently in primary human tumors. These mutations most commonly affect codons 12, 13 and 61 by single-base substitutions, producing altered forms of the $p21^{ras}$ proteins that are insensitive to GAP-regulated hydrolyzing activity, thus representing constitutively activated proteins that elicit mitogenic signals independently of ligand stimulation.

Following activation, and thus formation of the active Ras-GTP complexes, $p21^{ras}$ transduces its signals through specific effectors. Among the various effectors identified to mediate *ras* signals, Raf and PI3K are considered the most important and best characterized. In part, effector activation proceeds through its translocation into the cell membrane where it induces the initiation of a series of signaling events (Fig. 4). Raf activates the MEK1/2 dual-specificity serine/threonine kinase, which in turn activates, again by phosphorylation, specific mitogen-activated protein kinases (MAPKs) that finally, following translocation into the nucleus, induce the phosphorylation and thus activation of various transcription factors. Alternatively, PI3K activation results in the conversion of phosphatidylinositol-4,5-phosphate (PIP_2) to phospatidylinositol-3,4,5-phosphate (PIP_3). The elevation of PIP_3 results in the stimulation of Akt/protein kinase B (PKB) serine/threonine kinase activity that induces the phosphorylation of downstream targets. Finally, this cascade, among its other effects, results in the inhibition of proapoptotic cellular cues, e.g. by suppression of Bad-mediated inhibition of antiapoptotic proteins such as Bcl-X_L, inhibition of the apoptosis-related protease caspase-9 and activation of NF-κB levels by suppressing the activity of its inhibitor IκB.

Apparently, signaling through *ras* activation is subject to a considerable degree of amplification since from one activated Ras protein molecule, various different effectors are activated that, in turn, induce the activation of even more downstream factors by phosphorylation until the signal reaches the nucleus and is interpreted as a transcriptional activation event for the target genes. Furthermore, the wide variety of *ras* effector proteins is considered responsible for the pleiotropy of the *ras*-dependent effects, following the preferred use of one different effector molecule over another. This finds particular application when *ras* activation must be interpreted in different cellular contexts. For example, in fibroblasts, *ras*-induced senescence is controlled largely by Raf, whereas inhibition of *anoikis*

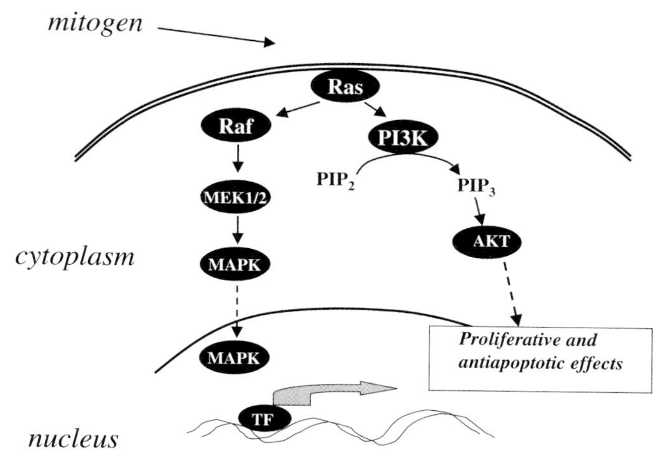

Fig. 4 The signaling cascade triggered by Ras operates through two major effectors, Raf and PI3K. Raf activates the MEK1/2 dual-specificity serine/threonine kinase which, in turn, activates specific MAPKs that finally, following translocation into the nucleus, induce the phosphorylation and thus activation of various transcription factors. Alternatively, PI3K activation results in the conversion of PIP_2 to PIP_3. The latter stimulates Akt/PKB serine/threonine kinase activity that induces the phosphorylation and activation of various downstream targets. TF = transcription factors that become activated by this pathway.

(suspension-induced apoptosis) is thought to be mediated by *ras* activated in a PI3K-dependent manner. In other cases, a combination of effectors is used such as in the case of *ras*-stimulated G_1/S transition of the cell cycle that requires both Raf and PI3K, and others that stimulate cyclin D1 expression and E2F transcriptional activation. Whilst discussing the stimulation of cyclin D1 by activated *ras*, it is important to mention recent experiments in cyclin D1 mutant mice that were resistant to *ras*-elicited mammary carcinogenesis. This experiment underlines the significance of cyclin D1 in *ras*-elicited malignant transformation.

ErbB2 Receptor Tyrosine Kinases

Certain receptors can function as oncogenes such as the receptors with tyrosine kinase activity (receptor tyrosine kinases). Receptor tyrosine kinases are single-pass transmembrane proteins consisting of an extracellular portion that confers ligand-binding activity and an intracellular portion possessing kinase activity. Furthermore, the intracellular domain possesses a C-terminal autophosphorylation sequence that determines the downstream effector molecules that will be recruited upon receptor activation and thus the specific signaling cascade that will be triggered. Upon ligand binding, receptor tyrosine kinases typically homo- or heterodimerize and their kinase activity is stimulated, triggering a complex cascade of signaling events. The driving force for this dimerization is the higher stability of the complex containing the two molecules of receptors and the ligand.

Among several receptor tyrosine kinases, the ErbB family of receptors play an important role in carcinogenesis. The ErbB family consists of four receptors (ErbB1–4). Among them, ErbB2 is a (notably highly oncogenic) orphan receptor with (yet?) unknown ligand and ErbB3, that is devoid to catalytic activity, recruits effectors, implying some regulatory activity. Several different, but homologous, ligands bind to these receptors, all of them containing the characteristic epidermal growth factor (EGF)-like motif. However, the ligands that bind to these receptors are not specific to them. For example, ligands that have been identified to bind with high affinity and activate ErbB1 include EGF, transforming growth factor (TGF)-β and others that differ in their binding affinity for this receptor as well the pattern of expression they exhibit in adult tissues and in different developmental stages. The relative quantity of the various homo- and heterodimers of the receptors is essential for the specific effector proteins that will be phosphorylated and activated. Among these effectors, several proteins have been identified, many of them with proven direct oncogenic activity such as Src, c-Yes, c-Abl, PI3K, Ras and others.

The oncogenic activity of ErbB receptors is exemplified by their transforming abilities, especially of ErbB1 and 2 *in vitro* and *in vivo*. Overexpression of ErbBs has been reported in primary cancers, such as that of ErbB1 in nonsmall cell lung carcinoma and that of ErbB2 in breast cancers. In these and other malignancies, the high levels of ErbBs frequently correlate with poor prognosis for the patients. In the case of erbB2, overexpression is frequently the direct result of gene amplification.

ErbB2, despite not having a known ligand, is the preferred partner for heterodimerization with the other receptor molecules because among other possible reasons it reduces the dissociation of the receptor from the ligand and elicits potent mitogenic signals, most likely conferring a proliferative advantage to the cells bearing it.

MicroRNAs as Oncogenes

That oncogenic activity can be produced only by certain proteins has been challenged recently after the discovery of some microRNAs with potential tumorigenic properties. MicroRNAs are short RNA oligonucleotides, usually 18–24 nucleotides in length, that are being produced by larger polycistronic transcripts following processing with specific RNases. Advances in molecular biology techniques, especially microarray technology, showed that such microRNAs (noteworthy more than 200 have been identified to date in humans) frequently exhibit aberrant expression in primary cancers. Interestingly, the pattern of expression of these microRNAs appears to predict quite precisely the aggressiveness of the tumors. Functional studies involving a microRNA designated mir-17-92, which is located in a region that is frequently amplified in lymphomas and other cancers, showed that this microRNA, and probably others as well, did indeed accelerate tumorigenesis in a mouse lymphoma model.

The oncogenic mechanism of microRNAs remains unknown; however, it is conceivable that they have implications in clinical research.

The Promise of Oncogenes in Applied Cancer Research

The discovery of oncogenes was a very promising advancement in cancer research as it provided the background for the development of rational cancer diagnostics and therapeutics. In particular, the identification of mutations in certain oncogenes that are associated with carcinogenesis in association with the development of novel molecular technologies renders the detection of the disease possible while it is still at a level that is asymptomatic. The importance of this approach lies in the fact that it targets the identification not of the actual tumor, but rather of its cause. Therefore, the disease will be detected at an earlier stage than that permitted by conventional approaches. For example, mutations in the *ras* oncogenes, the most common specific oncogene mutations in human tumors, can be identified if they are present in only few cells among a population of mutation-free cells in biological specimens such as saliva, sputum, etc. Furthermore, the association of specific mutations with certain subtypes of disease may facilitate the prediction of the prognosis of the disease more accurately than conventional clinical methods.

However, the most promising development is related to the therapy of cancer considering that mutations in oncogenes represent an early and causative alteration in carcinogenesis. As these oncogenic mutations must be expressed consistently for tumor maintenance (see Chapter 18), inhibition of the action of oncogenes represents a challenging target for cancer therapeutics. Indeed, various approaches have been followed that aim at the suppression of oncogene activity. For example, the HER-2/*neu* oncogene, which is frequently overexpressed in hormone-independent breast cancer with poor prognosis, can be targeted by an antibody that suppresses its activity. Indeed, such an antibody has been developed [trastuzumab (Herceptin)] and preliminary results of clinical trials have been quite promising.

ras oncogenes are also candidates for the development of anticancer therapies. Anti-*ras*-based therapy takes into consideration that a special post-translational modification of $p21^{ras}$, termed farnesylation, is necessary for the acquisition of full activity. Farnesylation is performed by some special enzymes termed farnesyltransferases that can be inhibited using exogenous agents. Clinical trials using such farnesyltransferase inhibitors are ongoing and show promising results.

Apart from the usual obstacles to the application of such therapies, e.g. toxicity, etc., a major methodological problem is that tumors frequently acquire additional mutations that render them independent of the initial oncogenic mutations. Thus, even complete suppression of these mutation does not warrant tumor suppression.

Bibliography

Alroy I, Yarden Y. The ErbB signaling network in embryogenesis and oncogenesis: signal diversification through combinatorial ligand–receptor interactions. *FEBS Lett* **1997**, *410*, 83–86.

Gasparini G, Longo R, Torino F, Morabito A. Therapy of breast cancer with molecular targeting agents. *Ann Oncol* **2005**, *16 (Suppl 4)*, 28–36.

He L, Thomson JM, Hemann MT, Hernando-Monge E, Mu D, Goodson S, Powers S, Cordon-Cardo C, Lowe SW, Hannon GJ, Hammond SM. A microRNA polycistron as a potential human oncogene. *Nature* **2005**, *435*, 828–833.

Lu J, Getz G, Miska EA, Alvarez-Saavedra E, Lamb J, Peck D, Sweet-Cordero A, Ebert BL, Mak RH, Ferrando AA, Downing JR, Jacks, T, Horvitz HR, Golub TR. MicroRNA expression profiles classify human cancers. *Nature* **2005**, *435*, 834–838.

Mitin N, Rossman KL, Der CJ. Signaling interplay in Ras superfamily function. *Curr Biol* **2005**, *15*, 563–574.

Nilsson JA, Cleveland JL. Myc pathways provoking cell suicide and cancer. *Oncogene* **2003**, *22*, 9007–9021.

Patel JH, Loboda AP, Showe MK, Showe LC, McMahon SB. Analysis of genomic targets reveals complex functions of MYC. *Nat Rev Cancer* **2004**, *4*, 562–568.

Pruitt K, Der CJ. Ras and Rho regulation of the cell cycle and oncogenesis. *Cancer Lett* **2001**, *171*, 1–10.

Secombe J, Pierce SB, Eisenman RN. Myc: a weapon of mass destruction. *Cell* **2004**, *117*, 153–156.

4
Tumor-suppressor Genes

Tumor-suppressor genes (TSGs), along with oncogenes, represent the best-studied genes associated with malignant transformation. These two classes of cancer-associated genes exemplify in the most characteristic manner the constant interplay between positive, tumor-promoting and negative, tumor-suppressive signals in the progression of the disease. In general, they encode for proteins that are localized in the nucleus and are involved, in a relatively direct manner, in the regulation of the cell cycle by providing signals or rendering the cell susceptible to such signals that inhibit cell cycle progression, or induce cell cycle arrest and apoptosis. Well known and widely studied TSGs are the p53, Rb, p16 and others. Of course, certain exceptions to the aforementioned overgeneralization regarding their function and localization exist, such as the PTEN (phosphatase and tensin homolog deleted on chromosome 10) TSG that encodes for a dual-specificity phosphatase, that is a protein and lipid phosphatase at the same time, and is responsible for removing phosphate groups from certain kinases and inactivating them. Another TSG that differs in terms of localization is the type II receptor for insulin-like growth factor (IGF) that is a membrane receptor.

Based on their negative role in the regulation of cell proliferation, their normal activity must be diminished or lost during carcinogenesis and it is exactly this property that is the basis for the development of anticancer therapies that aim at the restoration of their function. It has to be mentioned that the term TSG, formally speaking, should also cover genes involved in the maintenance of genome stability, such as the DNA repair proteins. As it will be mentioned Chapter 5, loss of their activity increases the possibility of the acquisition of mutations targeting genes with regulatory functions. Therefore, in order to define these genes separately from classic TSGs, the term "caretaker genes" has been introduced as opposed to the term "gatekeeper genes" that includes those related to the control of cell proliferation (classic TSGs). In this chapter we will only refer to the gatekeeper genes as TSGs, for historical reasons. It has to be emphasized, however, that some of the basic mechanistic aspects of their inactivation are also applicable to the caretaker genes.

Apart from the manifestation of the tumor-suppressive phenotype of the TSGs in genetic assays, important evidence regarding their operation as such is usually obtained from experiments involving the introduction and (over-)expression of these genes in cells that result in the inhibition of cell proliferation *in vitro* and *in vivo*.

Understanding Carcinogenesis. Hippokratis Kiaris

ISBN 3-527-31486-5

Below, we will describe in some detail the genetic mechanism of inactivation of TSGs in carcinogenesis, and then mention some examples of important and commonly altered TSGs in primary human tumors.

Cancer Genetics of TSGs

Knudson's Two-hit Mechanism for Inactivation of TSGs: Basic Aspects and Applications

Studies on the genetic behavior of TSGs have been dominated by Knudson's two-hit mechanism for their inactivation that unified the involvement of TSGs in both familiar and sporadic cancers in a simple genetic model. This model was originally proposed for retinoblastoma, by comparing the hereditary with the non-hereditary form of the disease. Subsequently, Knudson's two-hit hypothesis was found applicable to (almost) all TSGs. According to this hypothesis, in order to achieve cancer progression, both normal (wild-type) alleles of a given TSG must be inactivated, thus requiring two genetic hits for loss of their function. In hereditary types of disease, one allele has already been inactivated by deletion, mutation or any other genetic mechanism that has occurred in cells of the germline. Thus, the other allele remains to be inactivated by similar or epigenetic mechanisms (see Chapter 7) (Figs. 1 and 2). Hereditary cancers progress faster than the corresponding sporadic cancers because the first hit has already occurred (i.e. has been inherited) and thus, the second hit alone is sufficient for the loss of activity of the TSG in question. Obviously, this single hit that operates as the second hit is more likely to occur than both hits together, as is needed in sporadic cancers. Actually, one of the major evolutionary advantages conferred by diploidy is thought to be exactly this protection of the harmful consequences of mutations that may directly produce phenotypes, as in haploid organisms that are more vulnerable.

Therefore, according to this model, the first hit can be a missense mutation or a small deletion that results in the inactivation of one allele, while the second hit results in the functional inactivation of the remaining allele and may occur by various mechanisms. Deletions of whole chromosomes or chromosome stretches are thought to provide such hits during carcinogenesis by reducing the haplotype of TSGs to hemizygosity. However, physical loss of chromosomes or chromosome fragments is not the only mechanism that accounts for the silencing of tumor suppressors. Apart from inactivating point mutations, additional phenomena may involve gene silencing by epigenetic mechanisms, such as promoter hypermethylation, or mitotic recombination that eliminates the wild-type allele and replaces it with a copy of the mutant allele (Fig. 2). Thus, all these alterations have been included under the general term "loss of heterozygosity" (LOH), which covers all mechanisms that functionally result in the abolishment of the heterozygous state of the TSGs and in the generation of cells bearing only mutant allele(s) either in the homozygous or hemizygous state, depending on the molecular

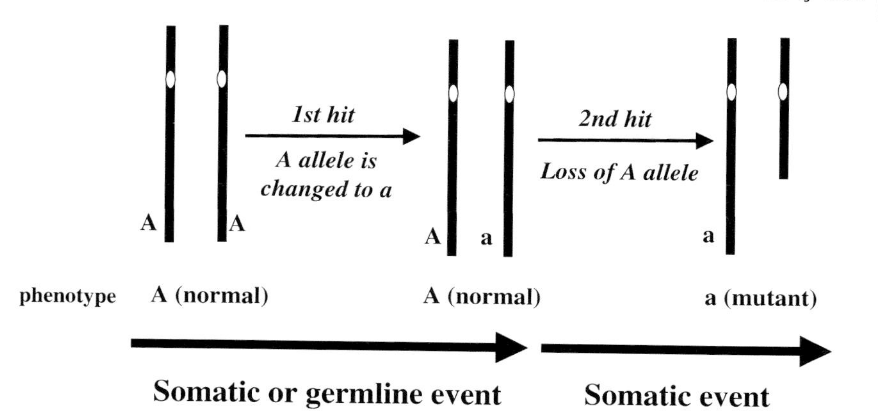

Fig. 1 Simplified overview of Knudson's "two-hit mechanism" for the inactivation of TSGs. Initially, a mutation occurs in one allele of a TSG. This mutation, that usually inactivates the specific allele, can happen in somatic cells or may have been inherited from the germline. Following this event, the cells have only one functional copy of the corresponding TSG. Subsequently, a second hit occurs in the remaining allele that leaves the cell without copies of this gene. "A" and "a" correspond to wild-type and mutant alleles, respectively.

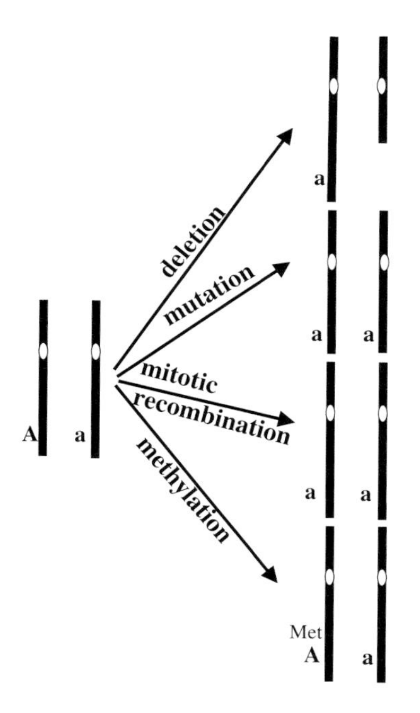

Fig. 2 Various mechanisms may account for the complete inactivation of a TSG during the second hit in Knudson's two-hit mechanism. For example, the remaining wild-type allele can be deleted, subjected to a second mutation or mitotic recombination, or, finally, epigenetically by becoming hypermethylated and therefore silenced.

mechanism that generated them (Fig. 2). The "bottom line", however, is that the genetic alterations must result in the loss of the corresponding gene's normal function.

LOH Analysis

The fact that LOH represents a landmark property implicating TSGs in carcinogenesis led several investigators to use this specific alteration in order to locate and subsequently identify new TSGs. These studies take advantage of the existence of several polymorphic markers that are widespread in the genome. These markers represent either restriction fragment length polymorphisms (RFLPs) due to a single base substitution or microsatellite polymorphisms due to different repetition numbers of the core microsatellite units (Fig. 3). In both cases the existence of polymorphisms renders possible the distinction of the two alleles and thus the identification of the potential deletion of one of them in the case that LOH has occurred. Usually, in studies such as those involving

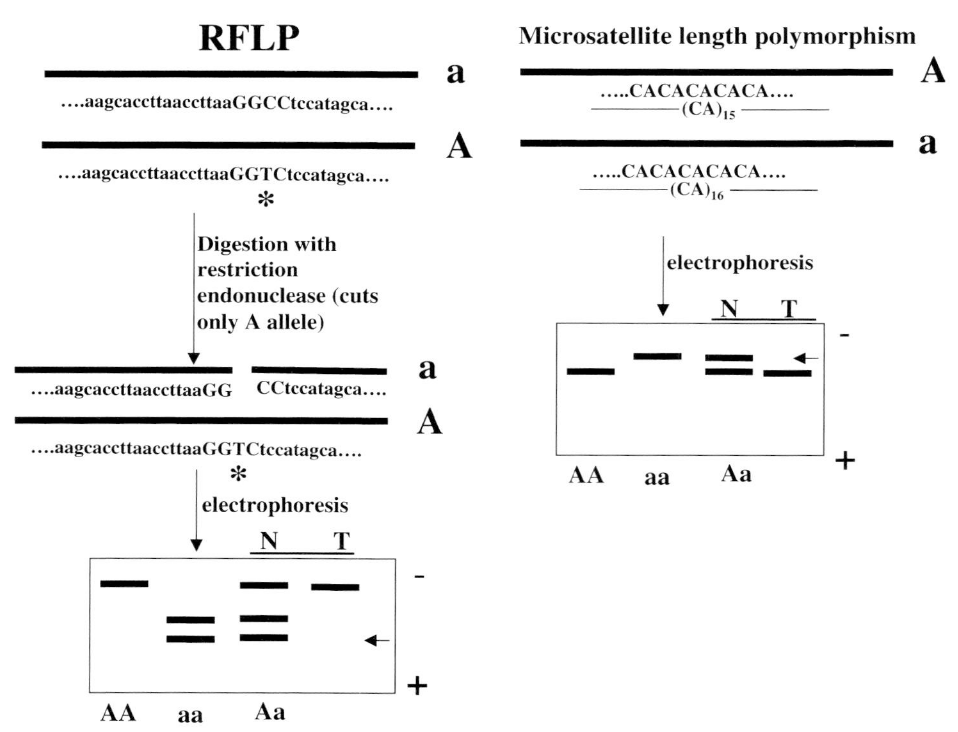

Fig. 3 RFLPs and microsatellite length polymorphisms constitute the basis for molecular genetics analyses that permit the mapping of TSGs. In both cases, following specific steps of manipulation of the sample DNA (usually involving polymerase chain reaction-based amplification of a defined DNA sequence), electrophoresis of the specimens permits the identification of the different alleles. In RFLPs, the distinction is based on the alteration of a sequence that corresponds to a site recognized by a restriction endonuclease; in microsatellite length polymorphisms, the differences are due to the different numbers of repetitions of the core microsatellite unit (usually CA repeats for dinucleotides). The occurrence of LOH is reflected by the absence of one allele in the tumor DNA (the "a" allele indicated by the arrows). It is self-evident that only heterozygous samples are informative in such analyses.

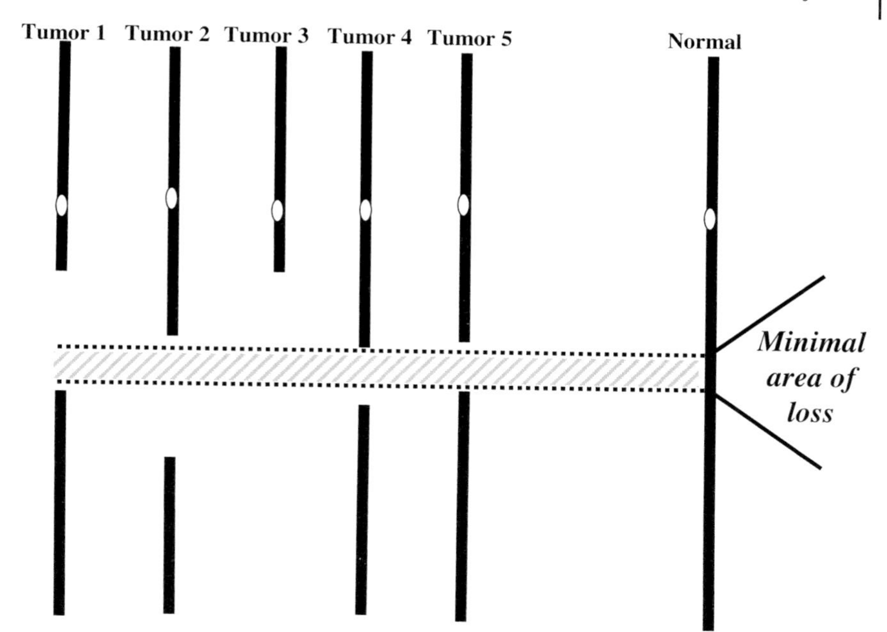

Fig. 4 Minimal area of loss. In this hypothetical example five tumor specimens exhibit LOH at overlapping regions in the q arm. The minimal area of loss is flanked by the deletions in tumors 4 and 5. For simplicity only one normal chromosome is shown; however, in reality, matched tumor and normal DNAs are used in these analyses. It is apparent that LOH analyses should involve many samples and the more informative specimens are those with relatively short regions with LOH.

LOH analysis in order to identify the location of novel TSGs, a bank of specimens is analyzed using such polymorphic markers. In the case that a candidate TSG is involved in the development of the disease, overlapping chromosomal fragments containing this TSG(s) should have been deleted. The commonly deleted area of different specimens, usually designated as the "minimal area of loss", is the portion of the chromosome that displays LOH at a higher incidence between different tumor specimens and thus it is likely to contain the candidate TSG. Shortening (narrowing down) this area by using markers at high density (many markers located closely to each other) renders the positional cloning of this gene possible (Fig. 4). Consistent with Knudson's two-hit hypothesis, a strong genetic confirmation that the candidate TSG indeed operates as such is the identification of inactivating mutations in the remaining allele.

However, the performance of LOH analysis in order to identify TSGs on the basis of their genetic behavior has certain methodological and conceptual restrictions. Such limitations probably explain the limited number of LOH analyses that have reached final success, as reflected in the cloning of TSGs, despite the enormous efforts undertaken by various investigators.

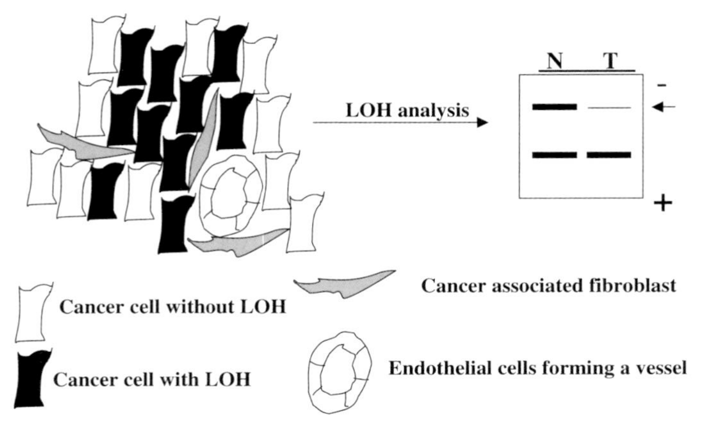

Fig. 5 Tumor heterogeneity affects LOH analysis. In practice, almost always, LOH analyses display the pattern shown in the right panel, with the "deleted" allele never being completely absent from the tumor DNA. The source of this amplified allele is DNA from cancer-associated fibroblasts, endothelial cells of the tumor vasculature as well as from cancer cells without LOH in the marker that is under investigation.

The methodological limitations are predominantly due to the fact that the detection of LOH presumes that the analysis is performed in homogenous cell populations that have lost the allele of interest in a uniform manner. This is of course not the case for two reasons. Tumor specimens never consist exclusively of cancer cells, but also contain, frequently at considerable levels, stromal cells that are genetically (almost) normal (see also corresponding chapter on tumor stroma). In addition, the same cancer cells exhibit significant heterogeneity due to clonal development of tumors. Alterations that have occurred in all cancer cells are in principle only those that have occurred very early in the development of the disease, while other deletions that have occurred at later stages will be present only in a subset of the cancer cells (Fig. 5). Therefore, in practice, LOH that results in the complete disappearance of the deleted allele only seldom (if ever) occurs, while in most cases what is apparent is the reduction in the intensity of the band (or signal in general, depending on the detection methodology) that corresponds to the deleted allele (Fig. 5). Investigators usually set some arbitrarily limits beyond which the LOH is scored as positive. Furthermore, and for the abovementioned reasons, homozygous losses that provide very strong evidence for the presence of TSGs in the deleted area cannot be easily identified with LOH analyses.

Conceptual restrictions are due to the oversimplification of Knudson's two-hit model for the mechanism of inactivation of TSGs in carcinogenesis. While this mechanism is indeed applicable in many cases, especially those involving cancer predisposition syndromes, it is based on the assumption that in sporadic cancers the affected locus is in the diploid condition when the first hit occurs and remains

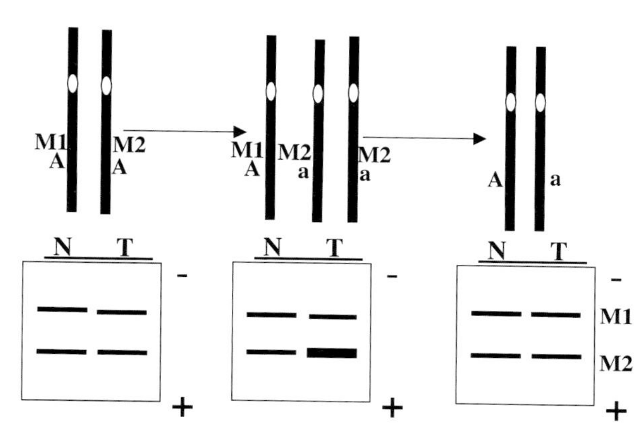

Fig. 6 Chromosomal aberrations affect the consequences of LOH and the outcome of the corresponding analysis. In this hypothetical example LOH occurs in a triploid tumor and results in the generation of a haplotype that is similar to that of the normal cells. "A" and "a" indicate normal and mutant alleles of the TSG, while "M1" and "M2" are two different alleles of an adjacent microsatellite marker.

as such until the second hit. However, in reality, tumors are polyploid or aneuploid – a state that strongly affects the exact process and the outcome of LOH. Strictly speaking, the genetic events of aneuploidy and polyploidy constitutes a LOH event in terms of the abolishment of the heterozygous state. For example, assuming that a TSG is located in an amplified (i.e. triploid) chromosomal region, then the first hit will result in a state where one wild-type and two mutant alleles exist. LOH in that case should result in the heterozygosity of the corresponding locus (Fig. 6). Alternatively, in the case that a mutation occurs before the genetic event generating triploidy, the latter doubles the dose of the mutant allele as compared to the normal.

Advances in molecular technologies, such as comparative genomic hybridization and DNA chips, in combination with developments in microdissection techniques that may isolate precise and defined cell populations within the tumor, may solve some of these problems, especially those related to the methodological limitations of LOH analyses.

In addition to LOH, other mechanisms by which mutations in TSGs may contribute to carcinogenesis involve the generation of dominant-negative forms of the TSG that have the ability to inactivate the remaining wild-type allele, usually by binding or competition for specific substrates. In this case, at the functional level, it is anticipated that the presence of dominant-negative mutations mimics the phenotype of homozygous deletions. Alternatively, a specific mutation in the TSG may not interfere with the activity of the wild-type allele, but rather generate a neomorphic allele that has acquired novel functions. Finally, as will be described below in more detail, a single hit resulting in the loss of function of a given TSG may be sufficient for tumorigenesis if this is associated with a reduction of the gene dosage and consequently with dose effects. An overview of these mechanisms by which mutations in TSGs can be associated with carcinogenesis is shown in Fig. 7. Importantly, the same gene can be the target of all these different mechanisms.

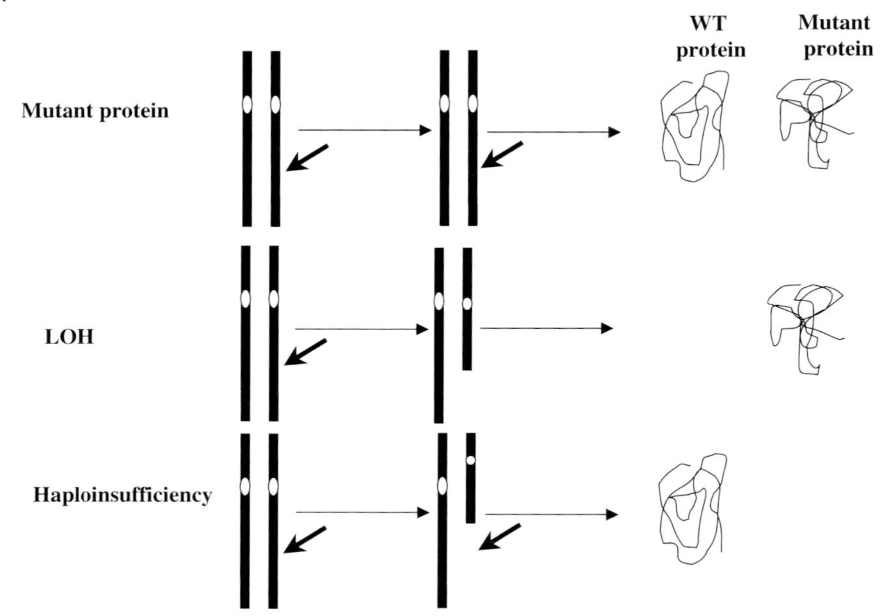

Fig. 7 TSGs can be altered in cancers by various mechanisms. For example, they can be subject to mutations resulting in the generation of neomorphic or dominant-negative alleles so that both wild-type (WT) and mutant forms of the protein are present, they can undergo LOH so that only the mutant form of the protein is being produced, or they can be haploinsufficient so the reduction in the dosage of the wild-type allele is sufficient to generate phenotypes. Frequently, any given TSG is involved in carcinogenesis with all possible mechanisms acting to varying extents.

Haploinsufficiency

According to Knudson's hypothesis that is applicable to the wide majority of TSGs, these genes are recessive in carcinogenesis in terms of the fact that mutations in the heterozygous state are not sufficient to produce phenotypic changes, especially cancer, unless the remaining allele becomes functionally inactivated. However, certain exceptions occur related to various TSGs that in the hemizygous state do not exert their tumor-suppressive effects sufficiently. These genes appear to operate in a manner that depends on the dosage, e.g. genes such as the cell cycle regulators of the CIP/KIP family, p21 and p27, respectively.

Formally, in order to prove that a TSG behaves as haploinsufficient in carcinogenesis, two important criteria must be fulfilled: (i) one must show that allelic deletions do occur in these genes and that they are associated with the carcinogenic process, and (ii) the remaining allele is expressed in cancers, notwithstanding at reduced levels due to the resulting hemizygosity, in its wild-type form without being affected by mutations.

It has to be noted that for almost all TSGs, reports on deletions without mutations at the remaining alleles are available for a subset of the cases studied, in both primary human and experimental animal tumors, implying that to some extent they may all bear haploinsufficient properties. The latter should not be surprising since the link between haploinsufficiency and gene dosage renders possible this behavior and is applicable to all TSGs.

Role and Properties of Some TSGs

Rb Pathway

Rb and p53 are nuclear proteins, and are considered among the most important regulators of the cell cycle. Rb was the first TSG cloned – a fact that dramatically changed cancer genetics. When mutated in humans it is responsible for the development of retinal tumors providing the genetic basis for Knudson to suggest his two-hit hypothesis for the action of tumor suppressors. Subsequently, Rb was found mutated in other types of cancer such as small cell lung carcinoma and osteosarcoma.

Oncoproteins encoded by tumor viruses frequently bind and inactivate Rb. Rb is responsible for G_1 arrest and prevents entrance into the S phase of the cell cycle by inactivating E2F transcription factors. E2Fs induce S phase transition by inducing genes that encode for S-phase cyclins and cyclin-dependent kinases (CDKs), DNA polymerase α, enzymes involved in nucleotide metabolism, and other proteins involved in DNA replication. This repressive action of Rb in gene expression is exerted at least in part through its ability to affect chromatin structure. However, the fact that more than 100 proteins have been reported to physically interact with Rb suggests that its role might be more complex than originally thought.

Inactivation of Rb by phosphorylation is mediated by specific CDKs, especially CDK4 and CDK6 that are under the negative regulation (inhibition) of the p16 TSG (Fig. 8). This represents an overview of the "Rb pathway" that is frequently

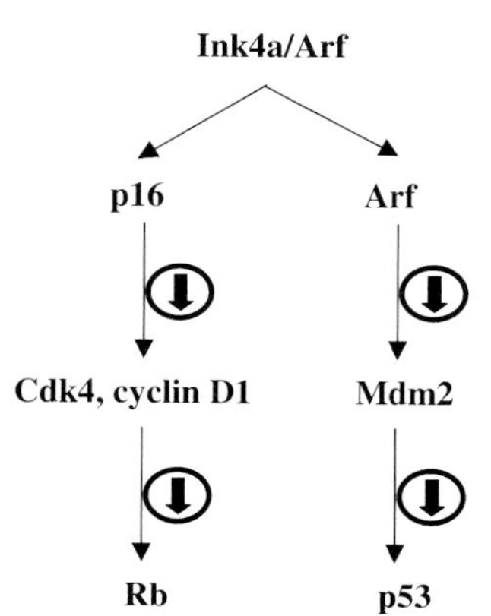

Fig. 8 The INK4a/Arf locus encodes for two TSGs that are transcribed by different promoters. These TSGs are $p16^{Ink4a}$ and $p19^{Arf}$ that inhibit the TSGs Rb and p53 through suppression of CDK4 and MDM2, respectively.

altered in many primary human cancers as evidenced by the relatively frequent detection of inactivating mutations in Rb or p16, and the overexpression of cyclin D1 and CDK4. p16, in particular, was originally identified in cases of familiar melanoma and it was frequently detected that it was subjected to homozygous deletions. Soon after its discovery it was thought to bear very promising clinical implications and was also termed MTS1 (multiple tumor suppressor 1).

p53 Regulatory Network

p53 is the central protein of another important regulatory network. In addition, it is mutated in almost 50% of primary human tumors, representing the most common genetic lesion in human cancer. It encodes for a 53-kDa transcription factor that is considered the major mediator of the response of the cell to stress triggered by DNA-damaging agents, hypoxia, activated oncogenes and other genotoxic stress-inducing factors. Its activation results in cell cycle arrest or apoptosis by binding onto specific regulatory regions in the promoters of target genes, causing induction or suppression of their activity. p53 is also the target of various post-translational modifications, importantly phosphorylation, that are important for its activation/inactivation through conformational changes that affect its specific binding to the DNA.

One of the pathways that regulate p53 activity is under the control of Arf1, the other product of the INK4α/Arf1 locus besides p16. Arf1 protein suppresses the activity of MDM2 protein, a negative regulator of p53 that notably is induced by activated p53, thus providing a negative feedback loop (Figs. 8 and 9). The suppressive role of MDM2 on p53 activity is produced by at least two mechanisms.

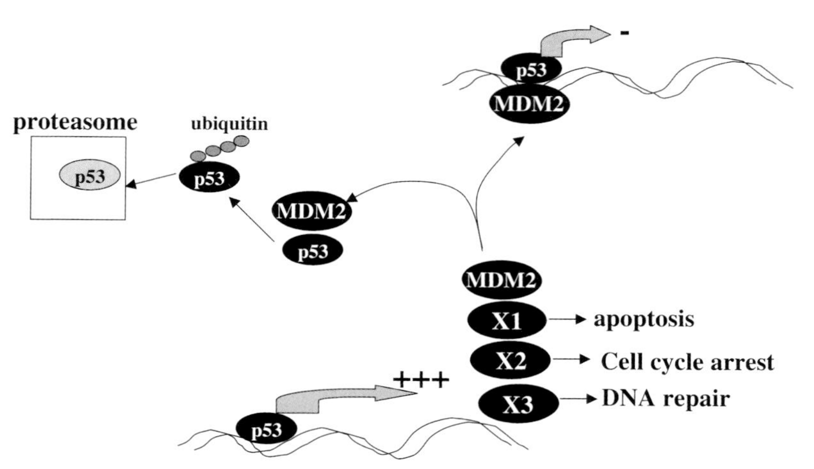

Fig. 9 The activity of p53 is regulated by MDM2 by at least two mechanisms: (1) MDM2 acts as an ubiquitin ligase targeting p53 to the proteosome and (2) binding of MDM2 to p53 alters its transactivating activity, thus modulating its ability to target downstream target genes.

It can bind the N-terminal transactivation domain of p53, thus preventing its interaction with other proteins that are critical for its action as a transcription factor. In addition, it operates as a p53-specific E3 ubiquitin ligase and therefore regulates its degradation. The importance of MDM2 in regulating p53 activity and thus affecting tumorigenesis is reflected by the fact that amplification of MDM2 is frequent in primary human tumors.

Whether following exposure of the cell to the appropriate signals, such as DNA damaging agents, it will respond by inducing growth arrest or apoptosis is highly dependent on p53 by mechanism(s) that still remain obscure. The tissue context appears to play a major role in determining the outcome of this response. For example, fibroblasts most of the times respond to DNA damage by p53-dependent cell cycle arrest, whereas epithelial cells respond by inducing apoptosis. Three models have been proposed to explain this decision, also called a "life or death decision". According to one model, p53 has the ability to sense and appropriately interpret the extent of DNA damage. Extended damage in the DNA induces strong and prolonged activation of p53 that will more likely induce apoptosis, whereas milder damage is more likely to elicit transient cell cycle arrest. According to an alternative model, the decision is predetermined in association with the particular cell type, which in turn makes a defined subset of gene promoters available to p53-dependent (*trans*)-activation by keeping them in transcriptionally active chromatin. Whether the corresponding genes are inducers of growth arrest or apoptosis determines the precise response of the cell via this p53-dependent mechanism. A third model suggests that the outcome of p53 activation is determined by the availability of the cofactors expressed under the given conditions. An example is offered by the interaction of p53 with modifiers of its activity, such as the members of the ASPP family and the p300-binding protein JMY that are thought to favor the p53-dependent transcriptional activation of proapoptotic genes, thus inducing apoptosis over growth arrest. Experimental evidence is available in support all three mechanisms.

The regulatory network in which p53 participates is very complex, as are the genes downstream that mediate its effects. One of these genes that represent targets of p53 is $p21^{Waf1}$, a member of the KIP/CIP family of CDK inhibitors, the activation of which is predominantly responsible for the cell growth arrest-inducing effects of p53. However, recent evidence showed that $p21^{Waf1}$ has also distinct p53-independent functions.

The induction of apoptosis by p53 is more complex because it can stimulate different proapoptotic genes. For example, it is capable of transactivating death receptors such as Fas/CD95/Apo-1, mitochondrial proteins such as Bax, Noxa, Puma, etc.

These functions of p53, i.e. induction of apoptosis or cell cycle arrest, are associated with its transactivation-dependent properties that involve its ability, operating as a transcription factor, to recognize specific sequences in the promoters of target genes and induce or suppress their expression. However, recent evidence showed that p53 can bind with high affinity to specific DNA structures regardless of the primary sequences, thus exerting a transactivation-independent function.

These structures include DNA mismatches and bulging DNA, providing strong evidence for the direct involvement of p53 in DNA repair. Among the five major mechanisms of DNA repair in eukaryotic cells (see Chapter 5), i.e. nucleotide excision repair, base excision repair, mismatch repair, nonhomologous end-joining and homologous recombination, the last involves almost exclusively p53's transactivation-independent function, while the others involve both transactivation-dependent and -independent p53 properties. In summary, p53 is thought to play a central regulatory role in determining whether in response to DNA damage the cell will repair the damaged DNA or will undergo cell cycle arrest and apoptosis.

Finally, it has to be mentioned that while the proapoptotic activity of p53 is being considered a "dogma" regarding its role in carcinogenesis, antiapoptotic effects of p53 have been suggested under certain conditions. $p21^{Waf1}$ is thought to play a role in this response that in certain cases, contrary its well-established anti-oncogenic action, reaches the level of contributing to certain oncogenic conditions.

In order to interpret these divergent and often controversial outcomes of p53, we have to keep in mind that formal descriptions of distinct activities, such as antiapoptotic and proapoptotic, represent oversimplifications devised for tutorial and descriptive purposes, and are tightly related to the experimental system used to obtain the results supporting these conclusions. In reality these effects are never dissociated from each other and the net outcome of different activities determines the mode of the response. For example, p53 activation that may induce cell cycle arrest under genotoxic stress can be considered as antiapoptotic in view of the fact that apoptosis has not been preferred over growth arrest. However, this is not equivalent to the classic antiapoptotic effects conferred by a "true" antiapoptotic gene such as Bcl-2.

CIP/KIP Family ($p21^{Waf1/Cip1}$ and $p27^{Kip1}$)

The CDK inhibitor p21 (formally $p21^{Waf1/Cip1}$) was originally identified as the first gene transcriptionally activated by p53 and was thought to represent a universal inhibitor of CDKs. However, recent evidence implies that its role is more complex than originally anticipated and it may elicit effects that are independent of p53 activation.

The regulation of cell cycle progression by p21 is exerted by its ability to bind and form complexes with cyclins and CDKs, and inhibit their activity – a response that as previously mentioned is originally triggered by p53. Despite the fact that this represents the best-studied action of p21, recent experimental results show that complexes between p21, cyclins and CDKs can be found in active forms. Furthermore, under certain conditions p21 is needed for G_1/S progression, thus implying a positive role in this process. Various mechanisms have been proposed to be important for these divergent effects of p21. It appears that the subcellular localization of p21 plays an important role in its effects – whereas nuclear localization is necessary for cell cycle inhibition, cytoplasmic localization of p21 may result in cell cycle progression and protection from apoptosis. The precise levels of p21 may also play a key role. Basal levels of p21 may play a positive role in the formation of the cyclin/CDK complexes and thus G_1/S progression, whereas increased p21 levels block CDK activity. Thus, it appears that progressively increasing levels of p21 transforms this protein from a tumor promoter to a tumor suppressor.

Apart from transcriptional activation, p21 is also regulated at the level of protein stability, being subjected to proteosome-dependent degradation that is regulated by the proliferating nuclear antigen (PCNA) – a protein that is commonly used as a marker of proliferating cells.

$p27^{Kip1}$ is the other member of the CIP/KIP family of CDK inhibitors that, in a similar manner to p21, binds to and inactivates CDKs and inhibits cell cycle progression. p27 is one of the first examples of a tumor suppressor clearly shown to function as haploinsufficient in carcinogenesis. Initial evidence came from studies on human tumors showing the reduced expression of p27 correlated with aggressiveness of the disease and the absence of mutations. Subsequently, conclusive results showing the haploinsufficiency of p27 were provided by experiments in mice indicating that reduction in the gene dosage of p27 accelerates tumorigenesis, whereas the remaining copy of the p27 allele is expressed and does not bear mutations.

Bibliography

Classon M, Harlow E. The retinoblastoma tumor suppressor in development and cancer. *Nat Rev Cancer* **2002**, *2*, 910–917.

Oren M. Decision making by p53: life, death and cancer. *Cell Death Differ* **2003**, *10*, 431–442.

Santarosa M, Ashworth A. Haploinsufficiency for tumour suppressor genes: when you don't need to go all the way. *Biochim Biophys Acta* **2004**, *1654*, 105–122.

Sengupta S, Harris CC. p53: traffic cap at the crossroads of DNA repair and recombination. *Nat Rev Mol Cell Biol* **2005**, *6*, 44–55.

Sherr CJ. The INK4a/Arf1 network in tumor suppression. *Nat Rev Mol Cell Biol* **2001**, *2*, 731–737.

Tomlinson IPM, Lambros MBK, Roylance RR. Loss of heterozygosity analysis: practically and conceptually flawed? *Genes Chromosomes Cancer* **2002**, *34*, 349–353.

Weinberg WC, Denning MF. $p21^{Waf1}$ control of epithelial cell cycle and cell fate. *Crit Rev Oral Biol Med* **2002**, *13*, 453–464.

Weiss RH. $p21^{Waf1/Cip1}$ as a therapeutic target in breast and other cancers. *Cancer Cell* **2003**, *4*, 425–429.

Zamzami N, Croemer G. p53 in apoptosis control: an introduction. *Biochem Biophys Res Commun* **2005**, *331*, 685–687.

Zhoua X, Raob NP, Colec SW, Mokf SC, Cheng Z, Wong DT. Progress in concurrent analysis of loss of heterozygosity and comparative genomic hybridization utilizing high density single nucleotide polymorphism arrays. *Cancer Genet Cytogenet* **2005**, *159*, 53–57.

5
Genomic Instability

It is continually stated throughout this book that carcinogenesis is a multistage process. This statement refers to both the physical history of individual tumors, i.e. that multiple genetic lesions are always detectable, and the level of the minimal genetic lesions required in order to cause the malignant conversion of normal cells. In other words, several genetic hits must collaborate to cause cancer, and following the establishment of malignancy additional genetic hits continuously occur and are subjected to the selection process, constantly altering the phenotype of the tumor.

During the early years of molecular cancer research the high level of genetic lesions that characterize the tumors led to theoretical considerations regarding how this increased mutational rate must have occurred. The fact that normal eukaryotic cells display a mutation rate of about 10^{-7} per nucleotide per cell division cannot explain the incidence of mutations seen in primary tumors. Thus, it has been proposed that tumor cells exhibit an increased mutational rate as compared to their normal counterparts, acquiring a "mutator" phenotype. This suggestion has been proved following (i) the identification of familial cancer syndromes that etiologically have been associated with the malfunction of mechanisms maintaining genomic integrity and (ii) the development of cancer-prone experimental animals due to various forms of DNA repair and chromosomal segregation defects. These forms of familial cancer include Xeroderma pigmentosum (XP), ataxia telangiectasia, Nijmegen breakage syndrome, hereditary nonpolyposis colorectal cancer (HNPCC), Bloom syndrome, etc. Furthermore, several sporadic cancers have also been found to bear somatic mutations in DNA repair genes, suggesting that the reduction in the fidelity of DNA replication is an intrinsic feature of all tumors, regardless of whether they are hereditary or sporadic.

Several molecular mechanisms involved in the maintenance of genomic stability have been implicated in carcinogenesis. They can be subdivided into two major categories, i.e. those protecting from chromosomal instability and those protecting from subchromosomal alterations involving even a single or a few bases only. They include nucleotide excision repair (NER), base excision repair (BER), mismatch repair (MMR), chromosomal segregation and others. The aim of this chapter is not to describe in detail the mechanisms contributing to the maintenance of genomic stability, but rather to underline how cancer and genomic instability are in-

Understanding Carcinogenesis. Hippokratis Kiaris

ISBN 3-527-31486-5

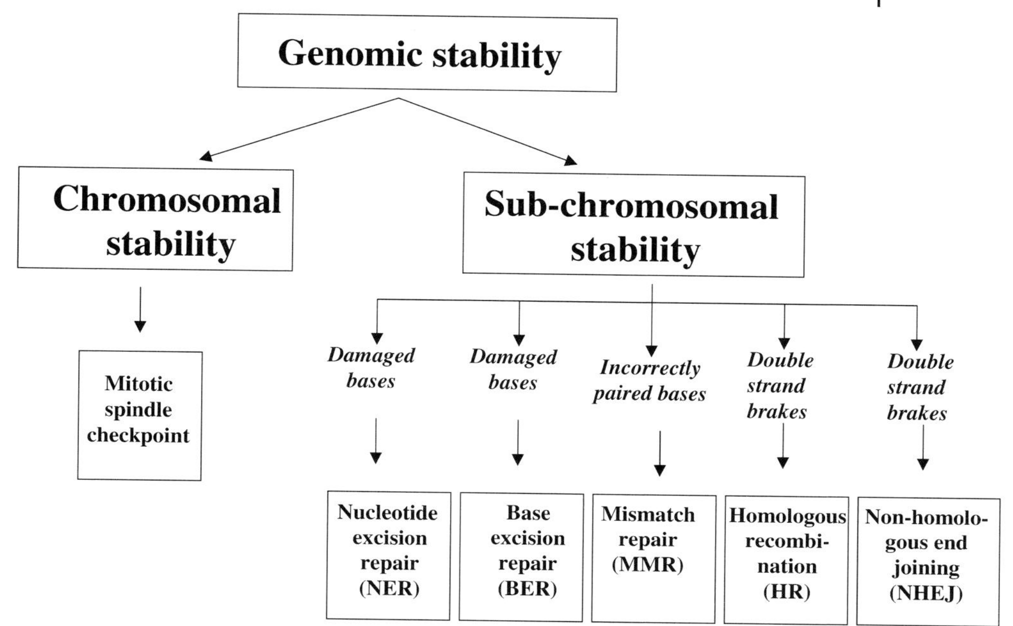

Fig. 1 Overview of the major DNA repair mechanisms responsible for the maintenance of genomic stability. Genetic stability is maintained at two different levels – the level of preserving the integrity of the chromosomal number and the level of maintaining the primary structure of DNA. Efficient correction mechanisms have evolved especially for the second (subchromosomal) level.

trinsically associated; however, some major characteristics of how the cell repairs damaged or inappropriate bases in the DNA are given below (Fig. 1).

An Overview of Mechanisms Preserving Genomic Integrity

In NER, damaged bases are recognized by a large multiprotein complex and removed from the DNA as part of oligonucleotides containing them, usually 25–30 nucleotides long. Subsequently, DNA synthesis operates using the normal DNA strand as a template in order to fill in the gap and the last nucleotide is ligated to the extant nucleotide. The vast majority of XP cases are due to defects in genes encoding for proteins implicated in NER. It is notable that at least seven complementation groups have been described corresponding to equal numbers of genes involved in NER. Complementation groups are genetically distinct entities implying the presence of individual genetic loci. A specific class of NER is called transcription-coupled NER and involves, in addition to other NER-related proteins, also BRCA1 and BRCA2 genes that have been associated with some forms of hereditary breast cancer.

In BER, DNA glycosylases, enzymes with different specificity recognizing individual or a defined subset of altered bases, play a central role. For example, when U is present in the DNA (while it should be present in the RNA only), a specific uracil-DNA glycosylase hydrolyses the *N*-glycosyl bond that links the uracil base to the deoxyribose phosphate, leaving a site at the DNA free of base. Subsequently, other enzymes repair these apurinic or apyrimidinic sites.

MMR recognizes nucleotides that are incorrectly paired and operates predominantly during DNA synthesis due to the limited fidelity of the DNA replication machinery. Members of the MSH, MLH and PMS families of proteins participate in multiprotein complexes that recognize loops due to nucleotide mismatches and repair them. A series of complex enzymatic reactions subsequently operates to repair these mistakes. Defects in MMR are considered responsible for a subset of familial and sporadic colon and other cancers, as will be described later in greater detail.

Homologous recombination (HR) and nonhomologous end-joining (NHEJ) refer to mechanisms that are employed when more serious DNA damage occurs, such as double-strand breaks (DSBs). In HR, the sister chromatid is (usually) used to guide repair; in NHEJ, the free ends are joined based on the homology of short repeats. Therefore, whereas HR is considered a repair mechanism that does not produce mistakes, NHEJ is considered an error-prone mechanism that repairs the DSB, but at the same time may generate small deletions or translocations. Following the generation of the DSBs, the sensor proteins that will recognize the defective DNA will subsequently dictate the specific repair pathway that will be followed. For example, binding of the Ku heterodimer results in the initiation of NHEJ, whereas binding of a protein complex termed Mre11–Rad50–NBS1 (NMR) is permissive for either NHEJ or HR. In this latter case, the factors that determine whether cells will undergo NHEJ or HR to repair the DSBs remain to be elucidated.

In general, it is thought that three distinct DNA damage checkpoints are functional in eukaryotic cells: the G_1/S checkpoint that prevents progression into the S phase and thus DNA replication if DNA damage has occurred, the intra-S checkpoint that corrects replication errors and lesions that escaped from the previous checkpoint, and the G_2/M checkpoint that prevents the initiation of mitosis in the presence of DNA damage. Before the activation and, indeed, even before the point of selection of the specific pathways that will correct the DNA lesions, the cell undergoes a transient cell cycle arrest. If the extent of DNA damage is "judged" to be catastrophic then apoptotic pathways are triggered that elicit cell death. A central role in these decisions of the cell in response to DNA damage is attributed to the p53 tumor suppressor that can function either in a transactivation-dependent manner to induce cell cycle arrest and/or apoptosis, or in a manner that can be both transactivation-dependent and -independent to induce the specific DNA repair pathways. This decision is based on various parameters such as the extent of DNA damage and the cell cycle conditions, with p53 playing a key role depending on its level and degree of post-translational modification. Under normal conditions and during limited DNA damage, p53 interacts with proteins involved in DNA repair and facilitates the correction of the DNA lesions. For extended damage, however, p53 is stabilized and induces apoptosis or cell cycle arrest by its transactivation-dependent function.

A Subset of Colon Cancers is a Prototype of Tumors Directly Associated with Increased Genomic (Microsatellite) Instability – MSI Tumors

Although historically this form of colorectal cancer, i.e. HNPCC, does not represent the first tumor type linked to DNA repair defects, its study has provided important clues about how DNA repair and carcinogenesis are intrinsically associated.

HNPCC is an autosomal dominant condition in which cancers arise from benign, probably adenomatous precursor lesions. Extensive molecular analyses revealed that HNPCC is characterized by two important features: (i) genetic linkage to loci associated with genes homologous to yeast genes encoding for DNA repair proteins (predominantly MLH1 and MSH2) and (ii) a high incidence of MSI in the tumor tissue (Fig. 2). Microsatellite DNA, due to its highly repetitive nature, and the lack of (apparent) selective pressure that favors the generation and stabilization in the genome of several alleles of different size (number of repetitions of the core unit) that do not affect the cell's viability, exhibits an increased mutational rate of about 10^{-5} per nucleotide per cell doubling. This is about 100 times higher

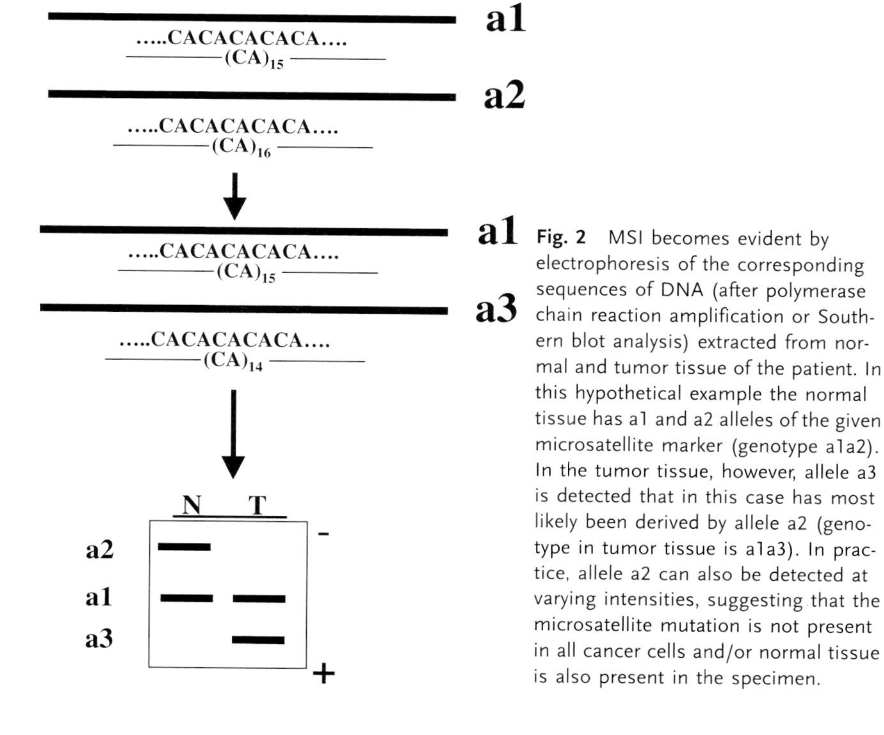

Fig. 2 MSI becomes evident by electrophoresis of the corresponding sequences of DNA (after polymerase chain reaction amplification or Southern blot analysis) extracted from normal and tumor tissue of the patient. In this hypothetical example the normal tissue has a1 and a2 alleles of the given microsatellite marker (genotype a1a2). In the tumor tissue, however, allele a3 is detected that in this case has most likely been derived by allele a2 (genotype in tumor tissue is a1a3). In practice, allele a2 can also be detected at varying intensities, suggesting that the microsatellite mutation is not present in all cancer cells and/or normal tissue is also present in the specimen.

than that of nonrepetitive DNA. During loss of heterozygosity (LOH) analyses aimed at identifying tumor-suppressor genes (TSGs) implicated in the development of the disease, investigators detected that tumor DNA from those malignant HNPCC lesions frequently bears microsatellite alleles that are absent from the normal tissue of the same patients. This observation is consistent with the acquisition of mutations at the microsatellite DNA of the cancer cells that are clonally expanded and become detectable. Thus, HNPCC tumors exhibit an increased mutational rate reflected by the incidence of MSI. It has to be emphasized that microsatellite mutations are neutral in the sense that they do not alter cell viability and they do not offer any proliferative advantage, as these sequences are noncoding and nonregulatory (with a few exceptions). This observation, in combination with the linkage of the disease to loci encoding apparent DNA repair genes, led to the suggestion that the etiologic factors responsible for HNPCC (and probably other types of cancer) are mutations in DNA repair enzymes. Interestingly, following the description of these mutations in specific colon cancers, a large number of reports rapidly confirmed and extended these findings to many different cancers, importantly sporadic cancers as well, suggesting that this phenomenon must have been noted by many investigators.

Colorectal cancers that progress through this pathway (MSI) usually harbor mutations in genes such as transforming growth factor-βRII, insulin-like growth factor-IIR, *bax*, BRAF and E2F4 while their karyotype is almost diploid.

Tumors Characterized By Chromosomal Instability – CIN Tumors

Another subset of colorectal cancers progresses through mechanisms involving CIN, resulting in polyploidy/aneuploidy, and are usually referred to as CIN tumors. Those cancers are usually free of MSI, and frequently bear mutations in genes such as APC, K-*ras*, SMAD4 and p53. Interestingly, XP which is due to defects in NER, is also a CIN(–) cancer. Although such classification constitutes an oversimplification, malignant tumors can be either aneuploid (CIN) or have defects in specific DNA repair pathways. In the latter case the gross chromosome number (and structure) is maintained. Thus, it is conceivable that CIN must induce the mutation rate by a manner similar (or equivalent) to DNA repair inactivation. Mechanistically, it is thought that aneuploidy is associated with tumorigenesis by accelerating LOH and reduction to homo-(hemi-)zygosity of TSGs or by increasing the aberrant signal of oncogenes undergone gene amplification.

The mitotic spindle checkpoint, which is responsible for assuring that chromosomes are normally aligned during metaphase and properly attached to the mitotic spindle before the separation of the chromosomes, is considered a frequent point of failure in CIN. Genes involved in the spindle checkpoint that have been identified as being altered in malignant tumors and patients suffering from cancer syndromes are hBUBR1, hMAD2, etc.

The impact of chromosomal instability in carcinogenesis has led to suggestions that contrary to mutations in oncogenes and TSGs, this (CIN) actually represent *the* initiating step in tumorigenesis that is able to activate (by amplification) oncogenes and destabilize (by allelic deletions) TSGs. Actually, it has even been proposed that the etiological association attributed to mutational events in carcinogenesis in certain instances may represent experimental artifacts. However, an alternative suggestion states that such genomic instability may have had occurred earlier in the cells, likely induced by telomere erosion, but subsequently the genome was stabilized, thus reducing its impact in the carcinogenic process.

Mitotic Recombination and Cancer: BLM Helicase

RECQ helicases in humans constitute a family of DNA repair proteins that also protect cells from genomic instability. They are unable to unwind blunt-ended duplex DNA, but they efficiently remove (unwind) Holliday junctions, G quadruplexes and other atypical secondary structures in the DNA. One of them, termed BLM, is responsible for Bloom's syndrome – a cancer predisposition syndrome. A diagnostic characteristic of the disease is the multifold increase in the frequency of sister chromatid exchange. Loss of the BLM gene results in abnormal DNA replication and increased recombination.

Chromosomal Rearrangements due to Telomere Dysfunction

As it will be discussed later in greater detail with regard to the role of telomerase in carcinogenesis (Chapter 8), telomeres protect cells from sister chromatid fusion and prolonged breakage/fusion/bridge cycles that eventually result in increased chromosomal instability. Normal cells avoid this phenomenon due to progressive telomere loss by entering a state termed replicative senescence. The breakage/fusion/bridge cycle progresses as follows (Fig. 3). During the replication of a chromosome that has lost its telomeres, sister chromatids are fused together at their ends. A bridge is then formed that breaks during anaphase and at the next round of DNA replication a new fusion is made followed by the subsequent steps of the breakage/fusion/bridge cycle. The resulting chromosomal instability is due to the fact that the break never occurs at the precise location of the fusion. Thus, depending on their relative position, duplications, deletions and other chromosomal alterations may be generated. Considering that these breakage/fusion/bridge cycles operate repeatedly, it is conceivable that they represent an efficient mechanism to explain chromosomal alterations commonly found in human cancer. In addition, during such cycles it is possible that while forming the bridge, the chromatids detach from one of the spindle poles resulting in the gain of an additional dicentric chromosome in one of the daughter cells and the loss of two chromosomes in the other daughter cell.

The fact that telomerase is active in the vast majority of human tumors, thus protecting telomeres from erosion and the aforementioned breakage/fusion/bridge cycles, may cast some doubt on how widespread this mechanism might

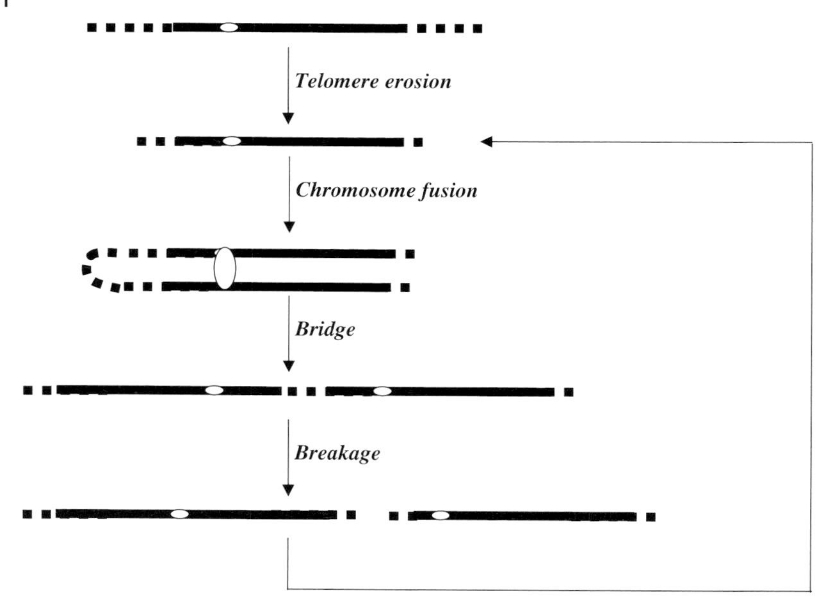

Fig. 3 The breakage/fusion/bridge cycle links telomere dysfunction with the induction of chromosomal instability as follows. During the replication of a chromosome that has lost its telomeres, sister chromatids are fused together at their ends. A bridge is then formed that breaks during anaphase and at the next round of DNA replication a new fusion is generated followed by the subsequent steps of the breakage/fusion/bridge cycle. The resulting chromosomes have various alterations depending on the relative position of the breakage.

be in primary tumors; however, it has been shown to operate in certain mouse and human tumors.

Clinical Value of Genomic Instability

Detection of MSI is considered a relatively easy and not particularly costly screening test. All that it requires is the isolation of some relatively good-quality DNA that is subsequently subjected to microsatellite analysis. International conventions have set criteria of whether a particular tumor should be considered as MSI-positive or -negative depending on the incidence of microsatellite markers affected. According to the Bethesda Guidelines, for example, each specimen should be subjected to analysis involving five particular microsatellite markers and according to the results is classified MSI-H(igh) if two or more loci are found positive, MSI-L(ow) when only one locus is positive or, finally, MS-S(table) if no evidence of MSI is detected. While this approach has importance in the classification of a tumor as MSI-positive or -negative, with certain prognostic implications (see below), the fact that the increased mutational rate as reflected in the MSI (notwithstanding at varying incidence) represents a feature of virtually all malignancies led several investigators to apply this observation to the early diagnosis of cancer. Therefore, biological fluids such as sputum, urine and other specimens that potentially contain cancer cells have been successfully used for the detection of MSI. In this approach the microsatellite pattern of DNA extracted from these specimens in which cancer cells might be present was compared to that of normal DNA, e.g. extracted from peripheral blood. The presence of novel microsatellite alleles absent from the paired normal control can be considered as a hint for malignancy.

A great advantage of detecting MSI as compared to other microsatellite alterations such as allelic imbalances, which might be present at even higher incidences, is that it is characterized by higher sensitivity. This aspect is of particular importance considering the heterogeneity of the cellular population examined that can be due not only to the heterogeneity of the cancer cells with respect to the particular genetic lesions they bear, but also (and more importantly) the unavoidable "contamination" of the cancer cells by normal tissue. Thus, an allelic imbalance can be masked much more effectively than MSI by normal tissue, providing false-negative results.

The detection of larger chromosomal alterations that belong to the category of CIN tumors has also been proposed to possess clinical value. The major obstacle to the wide applicability of this approach is that it requires more specialized laboratory equipment. For example, karyotype analysis, comparative genomic hybridization and fluorescent *in situ* hybridization (FISH) can be used very efficiently to detect specific manifestations of CIN, but their applicability, with probably the exception of FISH, is usually restricted in the clinical setting.

Bibliography

Diaz LA Jr. The current clinical value of genomic instability. *Semin Cancer Biol* **2005**, *15*, 67–71.

Friedberg EC. How nucleotide excision repair protects against cancer. *Nat Rev Cancer* **2001**, *1*, 22–33.

Gisselsson D, Höglund MD. Connecting mitotic instability and chromosome aberrations in cancer – can telomeres bridge the gap? *Semin Cancer Biol* **2005**, *15*, 13–23.

Murnane JP, Sabatier L. Chromosome rearrangements resulting from telomere dysfunction and their role in cancer. *BioEssays* **2004**, *26*, 1164–1174.

Rajagopalan H, Lengauer C. Aneuploidy and cancer. *Nature* **2004**, *432*, 438–441.

Risinger MA, Groden J. Crosslinks and crosstalk: human cancer syndromes and DNA repair defects. *Cancer Cell* **2004**, *6*, 539–545.

Scully R, Xie A. Is my end in my beginning: control of end resection and DSBR pathway "choice" by cyclin-dependent kinases. *Oncogene* **2005**, *24*, 2871–2876.

Sieber OM, Heinimann K, Tomlinson IPM. Genomic instability – the engine of tumorigenesis? *Nat Rev Cancer* **2003**, *3*, 701–708.

Siebera O, Heinimannb K, Tomlinson I. Genomic stability and tumorigenesis. *Semin Cancer Biol* **2005**, *15*, 61–66.

6
A Twist in the (Genetic) Tail: Cancer Epigenetics

That cancer is a genetic disease is mentioned repeatedly throughout this book. The meaning of this statement reflects the fact that alterations in the primary sequence of DNA at specific genes or regulatory regions affecting a few bases only or whole chromosomes, including gene dosage effects, are causatively associated with the onset of neoplasia. Apparently, those alterations, either quantitatively or qualitatively, deregulate the expression of specific genes that can cause malignancy. However, it appears that the expression of several loci can be altered through other mechanisms, most of the times reversibly, in a manner that can involve the modification of DNA by the addition or removal of methyl groups, loss of imprinting, and rearrangement and modification of the nucleosomes in chromatin. Such alterations have been termed epigenetic alterations and are thought to produce their effects by changing the accessibility of certain factors in the chromatin, thus changing the expression of the corresponding genes (Fig. 1).

By definition, epigenetic inheritance involves cellular information that is heritable, but is not carried in the DNA sequence. The study of epigenetic processes traces its origins to the fields of developmental biology and differentiation in which a key role has been identified.

Whilst originally defined as opposite, or better still, contradictory, cancer genetics and epigenetics are now in an era of synthesis undergoing strong interactions and interdependency. The latter is apparent considering that epigenetic changes carry a genetic basis, while the expression of certain genetic alterations notably depends on the status of specific epigenetic modifications. Thus, a given phenotype possesses both a genetic and epigenetic basis. Unification of the epigenetic and genetic bases of cancer will enable a clearer understanding of the disease.

Histone	*N-terminus modification*
H2A	**phosphorylation, acetylation**
H2B	**phosphorylation, acetylation**
H3	**phosphorylation, acetylation, methylation**
H4	**phosphorylation, acetylation, methylation**

Fig. 1 Histones are subjected to various types of modifications, such as acetylation, methylation and phosphorylation, that are determined by and also determine the status of epigenetic modifications in the DNA.

Understanding Carcinogenesis. Hippokratis Kiaris

ISBN 3-527-31486-5

Here we will describe some of these epigenetic processes and their association with cancer development, with the emphasis on the role of DNA methylation, the best-studied epigenetic mechanism for gene regulation, in carcinogenesis.

DNA Methylation

The study of DNA methylation represents the best-studied epigenetic mechanism for the regulation of gene expression. Since the early 1980s, reports on the pattern and levels of DNA methylation in normal and matched malignant tissue have established that differential methylation at specific loci characterizes virtually every type of cancer. Now we know that aberrant methylation is the most common alteration in human cancer, surpassing in incidence other events like DNA base substitutions and chromosomal alterations. The "net" degree of (over)-methylation favors the normal tissues since in general the quantity of 5-methylcytocine is reduced in malignancies, at about 20–60% as compared to its normal counterparts. It has been estimated that approximately 4–6% of all cytosines in the chromatin are methylated under normal conditions, corresponding to about 0.75–1% of all nucleotide bases. Hypomethylation at specific loci has also been reported, notably in benign lesions as well, suggesting that this alteration represents an early step in the carcinogenic process. Importantly, global demethylation in general increases with disease progression. However, a given fact that probably represents a noteworthy "paradox" is that hypermethylation at certain loci has also been demonstrated, complicating both the overall role and the precise underlying mechanism behind the deregulation of DNA methylation in malignancies (Fig. 2).

Recent progress in our understanding of the role of DNA methylation in carcinogenesis has been facilitated by the development of novel methodologies that allow the fast and accurate determination of the methylation status of certain DNA sequences. Traditionally, the methodology of choice to perform DNA methylation analyses has been the use of specific restriction endonucleases that are active

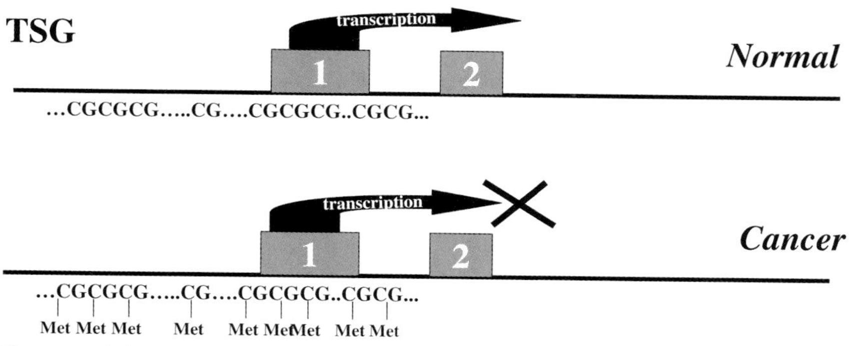

Fig. 2 Methylation in the promoter of TSGs suppresses transcription during carcinogenesis. This mechanism represents a major mechanism for gene silencing in malignancy.

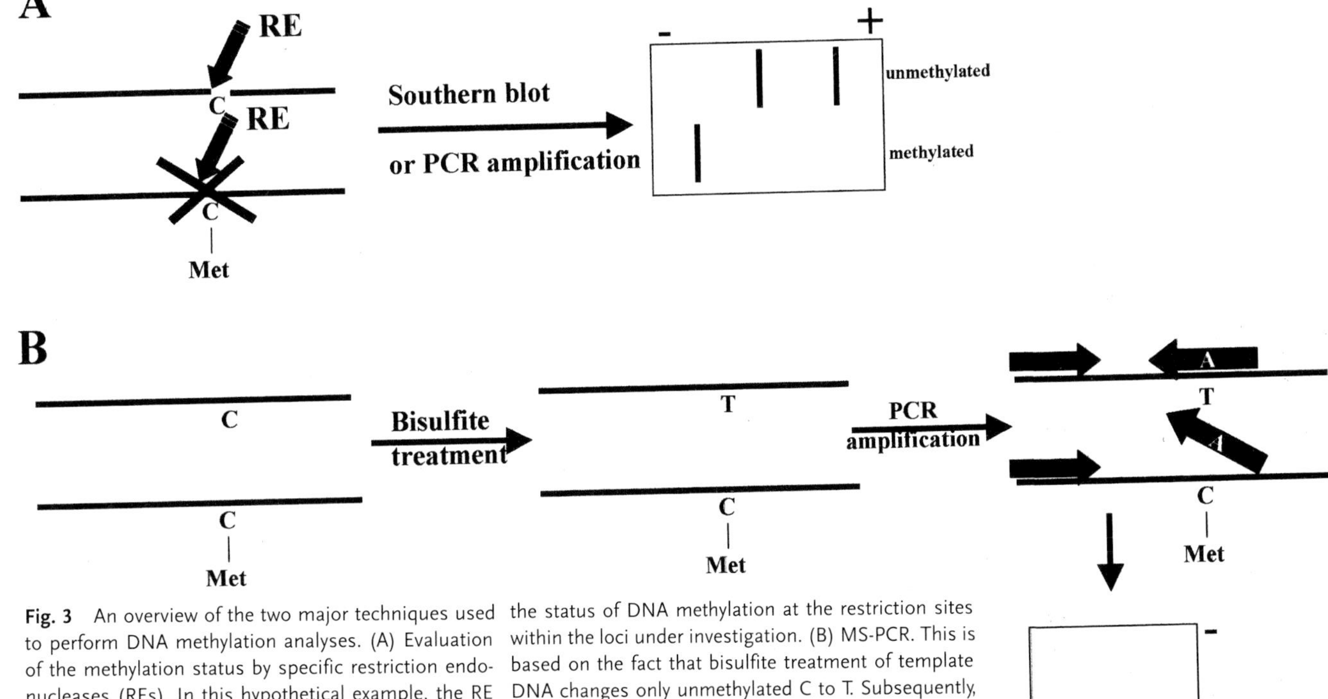

Fig. 3 An overview of the two major techniques used to perform DNA methylation analyses. (A) Evaluation of the methylation status by specific restriction endonucleases (REs). In this hypothetical example, the RE recognizes only the unmethylated DNA (upper), while the methylated DNA (lower) is resistant to the RE digestion. Subsequently, the digested DNA is subjected to Southern blot analysis or PCR amplification to detect the digestion products. Their pattern determines the status of DNA methylation at the restriction sites within the loci under investigation. (B) MS-PCR. This is based on the fact that bisulfite treatment of template DNA changes only unmethylated C to T. Subsequently, unmethylated and methylated DNAs have distinct primary sequences that can be amplified specifically by PCR following appropriate selection of primers. In this example only the unmethylated (T-containing) DNA is amplified.

only in methylated or demethylated sequences (Fig. 3). The major obstacle in such analysis is the requirement for sufficient, relatively good-quality DNA for Southern blot and the limitation that analysis is restricted by the existence of appropriate endonucleases that can recognize the sequence to be tested. A major advance was the development of the methylation-specific polymerase chain reaction (MS-PCR), which is based on the bisulfite treatment of template DNA that changes unmethylated C (only) to T (Fig. 3). Thus, following the chemical treatment, unmethylated DNA has a distinct primary sequence from the methylated DNA at the same locus, facilitating its specific PCR-based amplification. This approach, considering that it constitutes a PCR-based approach, has the great practical advantage that it can be applied even to archive material and biological fluids that in general do not offer DNA of the best quality and quantity. In a similar approach, bisulfite treatment can be coupled to other methods such as genomic sequencing that can facilitate the detection of novel genes subjected to differential methylation.

Other recent advances include the improvement of the immunohistochemical detection of met-C, that permits the evaluation of the extent of DNA methylation *in situ* and high-performance capillary electrophoresis advances that allow easy evaluation of the overall degree of DNA methylation.

How does DNA Methylation Occur and how is it Maintained?

Methylation occurs at the C5 position of cytosine at CpG dinucleotides and is achieved through the action of specific enzymes termed DNA methyltransferases (DNMTs). It evolved relatively late in evolution, and is thought to facilitate gene silencing and suppression of transposition of repetitive DNA sequences. CpG dinucleotides are not distributed equally through the genome, but exhibit higher concentrations in certain regions called CpG islands. Among these CpG islands, many are located within promoters and it is estimated that about 60% of human genes contain such CpG islands. Many mechanistic aspects on the role of methylation in cellular physiology have been derived from investigations addressing the role of X chromosome inactivation during development and differentiation.

The pattern of DNA methylation is set early during embryogenesis and subsequently maintained, in a regulated fashion, by the concerted action of the three DNMTs termed DNMT1, DNMT3A and DNMT3B. *De novo* DNMTs are predominantly DNMT3A and DNMT3B, while DNMT1 is a hemimethyltransferase (hemimethylase) that functions during DNA replication to maintain the pattern of DNA methylation (Tab. 1). This function of DNMT1 is due to the fact that it can recognize met-Cs on one strand of DNA in the replication fork, subsequently adding methyl groups into the corresponding C on the other strand. Another DNMT, termed DNMT2, has also been identified. Interestingly, as opposed to DNMT1 that is the only isolated by biochemical assay, DNMT2 has been identified by bioinformatic approaches due to its apparent homology with other methyltransferases. However, only some residual methyltransferase activity has been demonstrated *in vitro* and its precise role remains to be identified.

Tab. 1 The four major methyltransferases identified in humans and their properties.

Methyltrans-ferase	Properties
DNMT1	hemimethylase; maintains the methylation pattern during DNA replication
DNMT2	low methyltransferase activity; unknown role
DNMT3A	"true" methyltransferase; *de novo* methylation
DNMT3B	"true" methyltransferase; *de novo* methylation

Cancer and Aberrant DNA Methylation

In general, methylation results in the suppression of gene expression. However, in cancers, aberrant methylation can involve either the demethylation, and thus the overexpression, of oncogenes and the onset of genomic instability or the hypermethylation of tumor suppressors and therefore their repression. Alteration in the methylation pattern of human cancers has been reported in both directions. Apart from its direct effects on expression levels of certain genes, the altered methylation pattern, particularly hypermethylation, may silence DNA repair genes causing DNA repair deficiency with apparent consequences for the maintenance of DNA stability. Chromosomal instability has also been associated with altered DNA methylation. It has been demonstrated that in certain tumors, such as the Wilms' tumors, demethylation at satellite sequences in pericentric chromosomal regions may induce rearrangements that are not due to the induction of global DNA instability.

Globally, demethylation progresses with age in a tissue-dependent manner, thus providing an intriguing explanation for the increased incidence of cancer in the elderly. Collectively it is thought that demethylation generally increases genomic instability and contributes to loss of imprinting (see below).

A link between methylation status and cancer has recently been reported in studies evaluating methionine content in association with cancer risk. These studies showed that increased methionine content was related to a lower cancer incidence. This is in agreement with studies in rodents demonstrating that methionine-deficient diets induce the development of hepatocellular carcinoma without the administration of a carcinogen.

The molecular basis of global hypomethylation in cancer remains under investigation. Studies in liver cancer patients identified a common splice variant of DNMT3B which is associated with hypomethylation. Furthermore, experiments with mutant mice showed that deficiency of Lsh, a member of the SNF2 family that encode for ATP-dependent DNA helicases with a role in chromosome remodeling and maintenance of the normal DNA methylation pattern, results in a global defect of DNA methylation and genomic instability.

Genes that frequently undergo hypomethylation in various cancers include H-*ras*, cyclin D2, *bcl*-2, etc. These genes have a positive role in carcinogenesis that is stimulated by their overexpression, occasionally due to hypomethylation.

Despite the fact that globally chromatin hypomethylation is predominant in malignant tumors, hypermethylation characterizes specific loci. Early studies have shown that calcitonin, a marker for small cell lung cancer, and the Rb tumor-suppressor gene (TSG) were frequently hypermethylated in hereditary and sporadic cancers – an alteration associated with the reduced expression of these genes. Subsequent studies confirmed these findings and extended them to other tumor suppressors, such as INK4a that encodes for the p16 and Von Hippel-Lindau (VHL) syndrome gene, DNA repair genes, such as MLH1, and E-cadherin that is associated with invasion and metastasis. Interestingly, in many cases allele-specific hypermethylation in TSGs has been demonstrated, providing an alternative mechanism for their inactivation apart from allelic deletion. The importance of DNA hypermethylation at specific hotspots such as specific, CG-rich promoters as a cause for the silencing of TSGs is reflected in the fact that genome-wide searches have been devised to identify novel TSGs on the basis of their genomic structure – they frequently contain heavily methylated promoters in tumors. A similar approach is based on the treatment of cells with demethylating agents coupled with microarray analysis that aims to reveal the degree of overexpression.

With regard to the status of promoter methylation at CpG islands, cancers can be divided into two major categories – those characterized by frequent promoter methylation and those including the infrequent methylation groups. This observation led J-P Issa and coworkers to suggest the concept of the CpG island methylator phenotype (CIMP) in cancer.

The underlying mechanism that is responsible for targeted hypermethylation at specific loci remains unclear. It has been suggested that DNMTs, the enzymes responsible for the maintenance of the methylation pattern, preferentially bind to mismatched or damaged DNA. However, this explanation probably does not favor the concept that specific hypermethylation represents the initial mechanism by which tumor suppressors are inactivated, but rather the mechanism by which their inactivation due to methylation is maintained. Indeed, this is still a subject of debate that is supported by various experimental observations. For example, it has been shown that activation of MLH1 by 5-azaCdR is rapidly reversed spontaneously, implying that changes in the methylation state are secondary to other alterations.

An obvious question regarding methylation, however, is why this modification causes suppression of gene expression. Various models have been proposed to explain this observation. In certain cases it appears that methylation of DNA changes its conformation, thus preventing efficient recognition by DNA-binding proteins. This mechanism is tightly associated with the positioning of nucleosomes in the chromatin. In other cases, a more complex mechanism may be operational, related to the function of certain proteins that can recognize and "read" the methylated DNA. The action of these proteins is, in most cases, associated with transcriptional repression. Alternatively, DNA methylation may reduce the

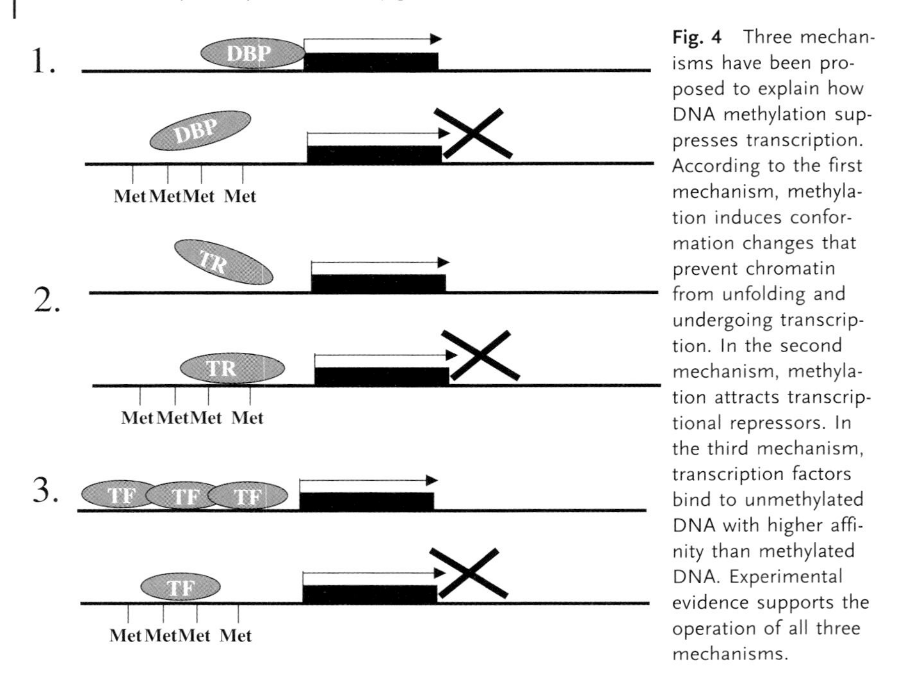

Fig. 4 Three mechanisms have been proposed to explain how DNA methylation suppresses transcription. According to the first mechanism, methylation induces conformation changes that prevent chromatin from unfolding and undergoing transcription. In the second mechanism, methylation attracts transcriptional repressors. In the third mechanism, transcription factors bind to unmethylated DNA with higher affinity than methylated DNA. Experimental evidence supports the operation of all three mechanisms.

binding affinity of certain transcription factors. Experimental evidence is available supporting all three mechanisms (Fig. 4).

One of the initial, if not the first, mechanistic findings explaining why methylated DNA was transcriptionally inactive was that MeCP2, a protein with high affinity for methylated DNA, recruits a histone deacetylase-containing complex and actively suppresses gene transcription. Following this observation, which favors the second mechanism detailed above, similar findings were subsequently reported.

In general, the modulated expression of genes with a regulatory role in the maintenance of DNA methylation, resulting in either hypermethylation or hypomethylation, is expected to simultaneously produce dual effects in carcinogenesis since it will act on both tumor suppressors and oncogenes at the same time with opposite consequences at the functional level. For example, global hypermethylation will cause the silencing of the tumor suppressors and the simultaneous activation of oncogenes, complicating our precise understanding of the role of methylation in carcinogenesis and the potential therapeutic implications it may have in the development of the disease.

Loss of Imprinting in Cancer

Imprinting refers to the phenomenon whereby certain genes are monoallelically expressed in a manner that only the maternal or paternal allele is active. Genes subjected to imprinting are insulin-like growth factor (IGF)-II, H19, BWS (Beckwith–Wiedemann syndrome), etc. The mechanism behind imprinting is methylation and, not surprisingly, imprinted chromosomal domains that contain several imprinted genes have been identified. Interestingly, these domains are conserved in different species. When loss of imprinting occurs in certain malignancies both alleles are expressed in the tumor tissue. For example, in Wilms' tumors biallelic expression of IGF-II represents a common alteration, which is related to the increased methylation of the H19 gene. Thus, the promotion of tumorigenesis by IGF-II appears to be dose-dependent and is attributed to its antiapoptotic properties.

Nucleosome Remodeling

The fact that DNA is coiled around histones forming the nucleosomes has implications for the accessibility of DNA to transcription factors and, thus, the degree of transcriptional activity. This pattern of nucleosome organization is a dynamically regulated process that is highly dependent on ATP. A central role in how nucleosomes are organized and, thus, which parts of chromatin will be transcriptionally active is played by the mode of modifications of histones, i.e. acetylation, methylation and phosphorylation. In general, histone deacetylation and Lys methylation result in gene silencing.

Despite the fact that histone modification and nucleosome organization have been noted for many years, the link with carcinogenesis has only recently become apparent. ALL1, the human homolog of the *Drosophila* gene trithorax with a role in regulating the expression of homeobox genes, is a tumor suppressor that is frequently altered in leukemias. It encodes for a protein that participates in a large protein complex that has chromatin remodeling, histone acetylation/deacetylation and histone methylation activities.

Variant Histones

Deposition of variant histones has also received a lot of attention. H2AX, the variant of histone H2A, is thought to mark double-strand breaks in DNA. Deletion of H2AX induces the development of lymphomas, suggesting that H2AX operates as a TSG. CENP-A is another variant histone that substitutes for H3 in centromeres. Analysis of primary colorectal tumors showed a correlation between overexpression of CENP-A and colorectal tumor development. Considering that only a few variant histones have been characterized in detail and that the human genome encodes for many more, an increasingly important role for these proteins in carcinogenesis will be most likely revealed.

A Dynamic Interplay Exists Between Different Epigenetic Mechanisms

Interestingly, crosstalk between different epigenetic processes becomes apparent by the colocalization of enzymes mediating DNA methylation and histone modifications – deacetylases, in particular. Even more intriguing, providing hints regarding the evolution of these mechanisms, is that in *Drosophila* the orthologs of the mammalian methyltransferases show transcriptional repressor function/histone deacetylase association instead of binding to met-C. This finding clearly suggests that these enzymes originally developed as proteins functionally associated with deacetylases and later acquired their methyltransferase activity that currently predominates.

In general, different epigenetic mechanisms exhibit strong synergy to regulate gene expression and act in a manner that is strongly cooperative. For example, it has been shown that DNA methylation may cause histone modification and *vice versa*, while nucleosome remodeling may induce DNA methylation (Fig. 5). However, the synergy between the various mechanisms of epigenetic DNA modifications takes many forms. It is conceivable that whereas their cooperativity, under physiological conditions, is responsible in maintaining genomic stability, under conditions where mistakes occur in one particular mechanism, those will accumulate and expand through other mechanisms as well.

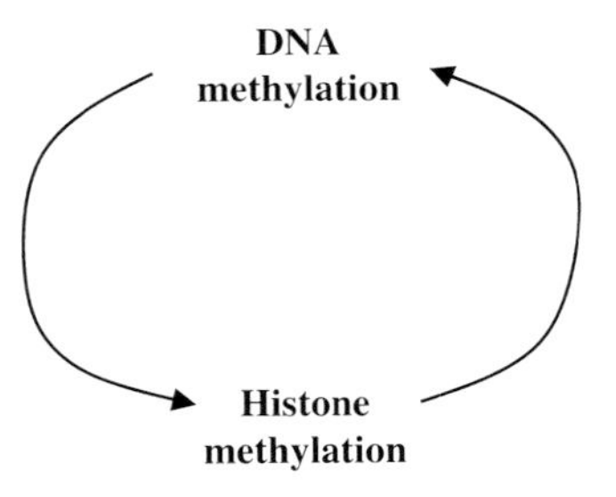

Fig. 5 Reciprocal functional interplay between DNA and histone modifications. DNA methylation promotes histone methylation that, in turn, promotes DNA methylation.

Translational Implications of Epigenetics in Cancer

Among the various epigenetic alterations that chromatin undergoes during neoplastic conversion, aberrant methylation is closer to the clinic. In general, the importance of methylation and the potential applicability in clinical practice is intrinsic to the fact that, contrary to the detection and evaluation of specific genetic lesions (with p53 and K-*ras* mutations being the most common), it reflects more general and wide alterations in the genome. Virtually all pathways with a role in carcinogenesis are affected by DNA methylation changes.

Detection of Cancer

Regarding the detection of cancer, it is considered possible that the disease can be detected with 100% success by using three or four specific methylation markers, (Esteller, 2005). Considering that new, sensitive and accurate methodologies have been developed, such as MS-PCR, then biological material (bronchoalveolar lavage, lymph nodes, sputum, urine, semen, ductal lavage and saliva) can be screened with a high success. In view of the fact that specific alterations in the methylation pattern of DNA occur early, such as the hypermethylation of p16 and APC, these screening assays may carry increased importance in the early detection of the disease.

Resistance to Therapy

The fact that that resistance/sensitivity to various modalities of anticancer therapy is regulated by the activity of certain genes implies that, at least for those where methylation represents an important epigenetic alteration, assessing their methylation status may predict the efficacy of anticancer therapy. Indeed, this hypothesis has already been proved for certain genes such as O^6-methylguanine DNA methyltransferase (MGMT), that encodes for a DNA repair enzyme involved in the correction of O^6-alkylguanine. Hypermethylation, and thus silencing, of MGMT is associated with a good response of glioma patients to drugs such as BCN (carmustine) temozolomide. Based on this principle, the acquisition of hormone independency by certain hormone-dependent neoplasms, with its apparent consequences for the therapeutic strategy to be followed, has been attributed at least in part to the silencing due to hypermethylation of the corresponding hormone receptors, including androgen, progesterone and, importantly, estrogen receptors.

Methylation-related Therapy

The early observations related to the hypermethylated status of specific loci at cancer cells rendered demethylation of chromatin an attractive therapeutic target. However, the intrinsic problem of such therapy, no matter how targeted and non-toxic it can be, is due to the fact that unavoidably demethylation will not target only TSGs, which constitute the desired targets for such treatment, but also oncogenes and growth factors. The loss of imprinting of IGF-II represents a good example of the undesired effects of demethylation therapy. Furthermore, DNA hypomethylation eventually results in increased genomic instability – another undesired consequence.

However, it has to be emphasized that in a similar manner to all conventional anticancer therapies, the overall efficacy is determined by the balance of its negative against its positive effects. Therefore, if the physiological consequences of the reactivation of TSGs are more potent than those from the reactivation of oncogenes, then it is conceivable that such anticancer therapy should be considered efficient.

It has to be noted, however, that certain experimental strategies are under development that aim at the specific activation of certain loci by taking advantage of gene therapy approaches. Their success should be seen in the near future.

Bibliography

Esteller M. Aberrant DNA methylation as a cancer-inducing mechanism. *Annu Rev Pharmacol Toxicol* **2005**, *45*, 629–656.

Feinberg AP, Tycko B. The history of cancer epigenetics. *Nat Rev Cancer* **2004**, *4*, 143–153.

Lund AH, van Lohuizen M. Epigenetics and cancer. *Genes Dev* **2004**, *18*, 2315–2335.

7
Nonautonomous Interactions in Carcinogenesis: Role of the Tumor Stroma

In primary tumors, cancer cells never grow alone by establishing contact exclusively with cells of the same type; rather, they form a very complex network of interactions with cells of different origin. These cells that constitute the microenvironment of the cancer cells are thought to represent the genetically normal component of a malignant tumor. They consist predominantly of the endothelial cells that form the microvessels, inflammatory cells related to the immune response of the organism against the developing tumor, and importantly fibroblasts (Fig. 1). These cell types constitute the tumor stroma. In certain cases, such as in some

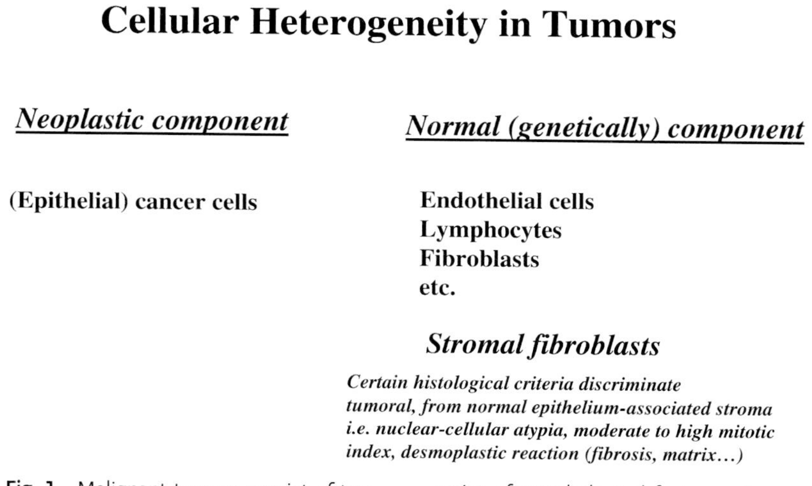

Cellular Heterogeneity in Tumors

Neoplastic component	***Normal (genetically) component***
(Epithelial) cancer cells	**Endothelial cells** **Lymphocytes** **Fibroblasts** **etc.**

Stromal fibroblasts

Certain histological criteria discriminate tumoral, from normal epithelium-associated stroma i.e. nuclear-cellular atypia, moderate to high mitotic index, desmoplastic reaction (fibrosis, matrix…)

Fig. 1 Malignant tumors consist of two components – the neoplastic component consisting of the cancer cells and the genetically normal component that contains the stromal cells, such as the fibroblasts, the endothelial and myoepithelial cells that are responsible for the tumor's vascularization, lymphocytes related to the immune response of the organism against the malignancy, etc. Stromal fibroblasts are of particular importance and they acquire a series of morphological features when associated with cancer cells, such as high atypia in the nucleus and the cell's morphology, increased mitotic index, and increased production of extracellular material (fibrosis, etc.) termed the desmoplastic reaction. The acquisition of these features by the stromal fibroblasts during carcinogenesis classifies them as a distinct entity and they are termed "cancer-associated fibroblasts".

Understanding Carcinogenesis. Hippokratis Kiaris

ISBN 3-527-31486-5

colonic, breast and pancreatic adenocarcinomas, the fraction of stromal cells and predominantly that of fibroblasts can reach up to 90% of the tumor mass, exemplifying the degree of stromal contribution in the tumor.

The study of stroma currently represents a very active field in tumor biology with promising implications in clinical practice. While the term stroma formally defines all the cellular and noncellular components of a tumor, excluding the malignant cells, in most cases it refers to the fibroblastic component only, as well as the extracellular material that is largely produced by the fibroblasts. These cancer-associated fibroblasts do not only respond to signals elicited by the cancer cells, but are also capable of affecting the morphology and behavior of the neoplastic component of the tumor, establishing a bilateral relationship with the cancer cells (Fig. 2).

The importance of stroma in tumor growth becomes apparent from a common observation, i.e. the rate of cell proliferation of a given cancer cell line *in vitro*, growing in a Petri dish, does not always coincide with the rate of growth of tumors formed by the same cells in appropriate experimental hosts such as immunoincompetent mice. Many cancer cells that grow very fast *in vitro*, when transplanted into animals exhibit a strikingly long latency for the development of palpable tumors and reduced cell proliferation rate as compared to others that while in the Petri dish grow relatively slowly, in mice form tumors rapidly that grow very fast. These contradictory observations regarding the *in vitro* against *in vivo* behavior of some cell lines are due, at least in part, to the fact that during *in vitro* culturing conditions the rate of cell proliferation essentially reflects the doubling capacity of the cells, whereas *in vivo*, in experimental animals, it is also determined by the availability and dependency of cells to nutrients, mechanical support requirements as well as sensitivity to contact. It is conceivable that during this transition from *in vitro* into *in vivo* growth conditions, a drastic change in the expression profile of the cells occurs which is responsible for their adaptation with the interaction with the stroma. How efficiently this is done most likely determines the growth profile of the cancer cells in experimental animals. In addition, the host that contributes the stromal component of a tumor in such experimental systems can dramatically affect the behavior of the cancer cells. Thus, the

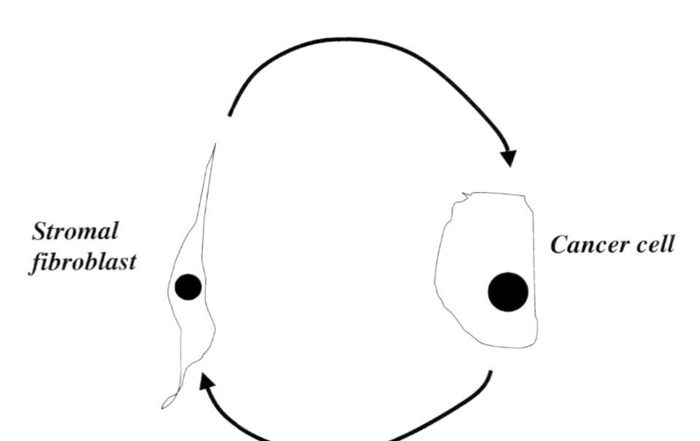

Fig. 2 Contrary to the conventional notion that stromal fibroblasts simply coexist with cancer cells in tumors, recent evidence attributed them a very important role in carcinogenesis as they receive and elicit signals by the cancer cells, promoting cancer growth.

tumorigenic profile of cancer cells is not the direct and sole consequence of their genotype, but is also affected by the properties of the cells that constitute their microenvironment, i.e. the tumor stroma.

Therefore, it is apparent that the study of cancer cells in culture, as well as the identification and subsequent evaluation of the role of the genetic alterations within the same cancer cells that constitute the malignant component of a tumor, represents an undoubtedly useful and informative approach; however, it possesses certain limitations representing a methodological abstraction.

The contribution of stromal fibroblasts in the tumorigenic process is exemplified by the fact that fibroblasts associated with carcinomas, but not tumor-free epithelium, can stimulate tumor progression of nontumorigenic epithelial cells, suggesting that they (fibroblasts) are sufficient to elicit potent oncogenic signals that, in turn, facilitate malignant transformation as manifested by the acquisition of tumorigenicity. In addition to the isolation of fibroblasts from cancer tissue, genetic manipulation of fibroblasts by hepatocyte growth factor (HGF) and transforming growth factor (TGF)-β (see below) overexpression, as well as irradiation of stromal fibroblasts, also promotes the malignant conversion of nontumorigenic mammary epithelial cells.

Although cancer-associated fibroblasts play a positive role in carcinogenesis, fibroblasts *per se* do not always promote tumor progression. In certain experimental systems the antiproliferative action of normal fibroblasts has also been demonstrated, implying an antitumor action. For example, *ras*-infected keratinocytes form tumors in mice only when injected alone or together with 3T3-immortalized fibroblasts, whereas injection with normal fibroblasts suppresses tumorigenicity. Consistent with these findings, normal fibroblasts exhibit antiproliferative activities on breast cancer cells in coculture experiments.

Tumor-associated Stroma has a Distinct Morphology: Desmoplastic Reaction

A few decades ago pathologists recognized that cancer-associated stroma is morphologically distinct from the stroma of normal tissue from the same origin. The set of these alterations is usually characterized by the term "desmoplastic reaction" and, among others, include increased mitotic index in the fibroblasts, atypia and certain changes in the extracellular matrix, such as increased production of specific collagens, fibronectin, glycosaminoglycans and proteoglycans.

Myofibroblasts represent the predominant type of fibroblasts in the tumor and can be identified by immunocytochemistry, by the expression of specific markers that include vimentin, α-smooth muscle actin, smooth muscle myosin heavy chain, desmin, calponin and α-integrin. It is noteworthy that this activation of a partial smooth muscle differentiation program, exemplified by the expression of the aforementioned proteins and resulting in the generation of myofibroblasts, has also been noted in the response to wound healing or inflammation, as myofibroblasts orchestrate the repair response. This is of particular importance because, as will be discussed later, an intrinsic link between cancer initiation and chronic inflammation exists which may involve the stromal fibroblasts. Note that inflammation also induces the formation of reactive stroma, which is morphologically indistinguishable from that triggered by malignancy.

Specific proteases, the matrix metalloproteinases (MMPs), secreted by cancer cells also play a pivotal role in the transition of stroma into its cancer-associated state, exemplified by stromelysin-1 (*str-1*) that when expressed ectopically by the mammary epithelial cells disorganizes the stroma in a manner that is sufficient to promote malignant conversion and breast cancer development.

Changes in the Expression Profile of Tumor Fibroblasts

Apart from the markers mentioned previously that are related more to their morphological characteristics, various other genes are differentially expressed in tumor-associated fibroblasts. Many of them involve growth factors that can act upon both the stromal and the cancer cells of a tumor. An example is offered by HGF that is expressed predominantly by the stromal fibroblasts, while its receptor c-Met is expressed by the cancer cells. It has been established that a paracrine stimulatory loop is operational in several cancers, mediated by HGF/c-Met. The implication of other growth factors, such as TGF-β, has also been established in the regulation of tumor–stroma interactions, but the mode of this interaction is more complex than that described for HGF/c-Met, because TGF-β and its receptors are expressed by both the cancer and the stromal cells, while its activation produces both tumor-inhibitory and -stimulatory results depending on the tissue and the experimental system used.

Tumor-associated Fibroblasts are not Necessarily Normal Genetically

Tumor-associated fibroblasts, contrary to cancer cells that continuously accumulate mutations, were thought to possess normal genetic material, which is free of mutations. Thus, any deviation from the normal morphology and properties for the tumor fibroblasts could be explained by changes in their expression profile, which apparently has been affected by the adjacent cancer cells. However, it appears that this is not always true – tumor-associated fibroblasts can also accumulate mutations that are notably similar to those identified in the neoplastic cells. Up to the present, mutations in the fibroblasts that have been described result in the inactivation of tumor suppressor genes (TSGs) such as p53 and PTEN (phosphatase and tensin homolog deleted on chromosome 10). Using p53 mutations in the fibroblastic stroma of mammary cancers as a model, and by performing tumor reconstitution experiments in which chimeric tumors were generated consisting of human mammary cancer cells and fibroblastic stroma with different genotypes, it was found that p53 mutations in fibroblasts confer more aggressive histopathology and an increased tumor growth rate than wild-type fibroblasts (Fig. 3). Such experiments underline the strong morphogenic properties of stro-

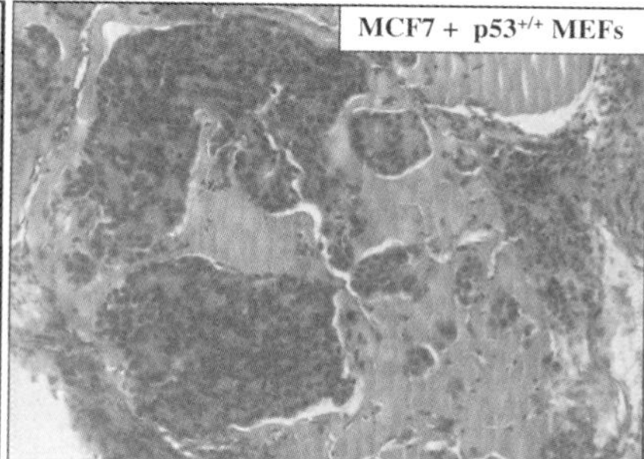

Fig. 3 An example of how the genetic background of stromal fibroblasts can affect the histopathology of tumors derived from the same cancer cells. In this particular example, MCF7 breast cancer cells were coinoculated into mice with fibroblasts that differed in p53 status. The genotype of fibroblasts is indicated in the top right corner of each microphotograph. (From Kiaris et al. (2005), reproduced with permission from *Cancer Res*).

mal fibroblasts and their impact in determining the properties of the resulting tumor. Interestingly, the mechanism of inactivation involves loss of heterozygosity (LOH) – the genetic alteration that is very common in the inactivation of TSGs in cancer cells. However, activation of oncogenes is also possible.

How alterations in the stromal fibroblasts integrate into the multistage process of carcinogenesis and especially how they interact with other processes that also exhibit nonautonomous mechanistic aspects, such as angiogenesis, remains unclear. Recently, Weinberg and coworkers provided evidence that activated fibroblasts (cancer-associated fibroblasts) possess a dual role in carcinogenesis: They promote tumor growth by secretion of growth factors such as the stromal-cell derived factor 1 (SDF)-1, that acts on cancer cells through its receptor termed CXCR4 (paracrine effect), and by recruiting circulating endothelial progenitor cells into the tumor, thus contributing to angiogenesis (endocrine effect).

Collective Remarks: "Modeling" the Tumor Stroma

Summarizing the aforementioned information, upon malignant conversion and following the initial proliferative cycles of a single cell that clonally is going to establish a tumor, stromal cells most likely play a negative role in carcinogenesis, protecting the organism from cancer by encapsulating the tumor and inhibiting its development. Subsequently, however, certain gene products of the cancer cells, such as growth factors and metalloproteinases, result in the modulation of the stroma that now becomes cancer-associated stroma. In turn, the stromal cells elicit signals that promote cancer growth, generating a cellular circuit subjected to positive feedback. Genetic and epigenetic changes targeting the stromal fibroblasts may also contribute substantially in the transition of stromal fibroblasts into this cancer-associated state (Fig. 4). During these early phases of carcinogen-

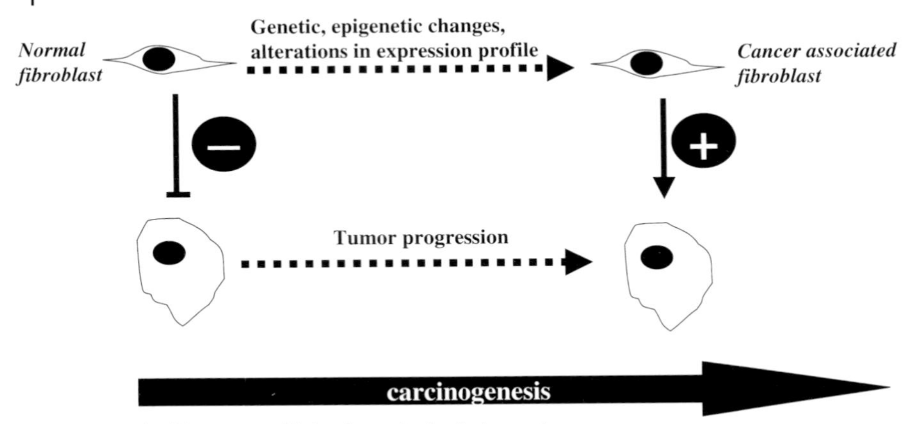

Fig. 4 Fibroblasts most likely play a dual role in carcinogenesis. During the initial stages of the disease they display a negative (inhibitory) role in tumor progression, whereas in subsequent stages they promote tumor growth.

esis, mutations in stromal cells (if present) are positively selected, adding to the heterogeneity of the stromal cells. The tumor is highly dependent on the stroma during this stage of cancer development – a feature that probably is abolished during subsequent stages.

The fact that malignant tumors, especially during the early stages of their development, exhibit a high dependency on tumor stroma renders therapy targeted to the modulation of tumor–stroma interactions a promising anticancer strategy. In association with the (in principle) lack of mutations in stromal fibroblasts, such an approach predicts a uniform response, devoid of heterogeneity due to differences in the genetic profile of the cells that are targeted.

Bibliography

Elenbaas B, Weinberg RA. Heterotypic signaling between epithelial tumor cells and fibroblasts in carcinoma formation. *Exp Cell Res* **2001**, *264*, 169–184.

Kiaris H, Chatzistamou I, Kalofoutis Ch, Koutselini H, Piperi C, Kalofoutis A. Tumor–stroma interactions in carcinogenesis: basic aspects and perspectives. *Mol Cell Biochem* **2004**, *261*, 117–122.

Kiaris H, Chatzistamou I, Trimis G, Frangou-Plemmenou M, Pafiti A, Kalofoutis A. Evidence for a non-autonomous role of p53 in carcinogenesis. *Cancer Res* **2005**, *65*, 1627–1630.

Kurose K, Gilley K, Matsumoto S, Watson PH, Zhou XP, Eng C. Frequent somatic mutations in PTEN and TP53 are mutually exclusive in the stroma of breast carcinomas. *Nat Genet* **2002**, *32*, 355–357.

Rowley D. What might a stromal response mean to prostate cancer progression? *Cancer Metastasis Rev* **2000**, *17*, 411–419.

Tlsty TD, Hein PW. Know thy neighbor: stromal cells can contribute oncogenic signals. *Curr Opin Genet Dev* **2001**, *11*, 54–59.

8
Telomerase and Cellular Immortality

A closer look at the semiconservative mechanism of DNA replication in eukaryotic cells predicts that gradually, during lagging strand synthesis, chromosomes will become shorter during each cycle of cell division – a prediction that has been termed the "end replication problem" (Fig. 1). Such progressive shortening of the chromosomes at their ends, under physiological conditions, triggers a cellular DNA damage response that is irreversible and does not depend on the availability of growth factors in the microenvironment of the cells. This state, which is regulated by p53 and Rb tumor-suppressor genes (TSGs), is called replicative senescence and represents the Hayflick limit. This "problem", i.e. bypassing replicative senescence due to telomere shortening, which should be faced by all normal cells after a certain number of divisions, has to be solved in specific cell types. The "end replication problem" is more evident in those cells that must exhibit the capacity for an indefinite lifespan, such as the cells of the germ line and, importantly, cancer cells, as well as other cellular populations such as embryonic and stem cells, activated lymphocytes, etc. Indeed, a specialized mechanism for the replication and extension of the ends of the chromosomes has been evolved that bypasses the consequences of the "end replication problem" and is based on the duplication of chromosome ends by a template-independent mechanism.

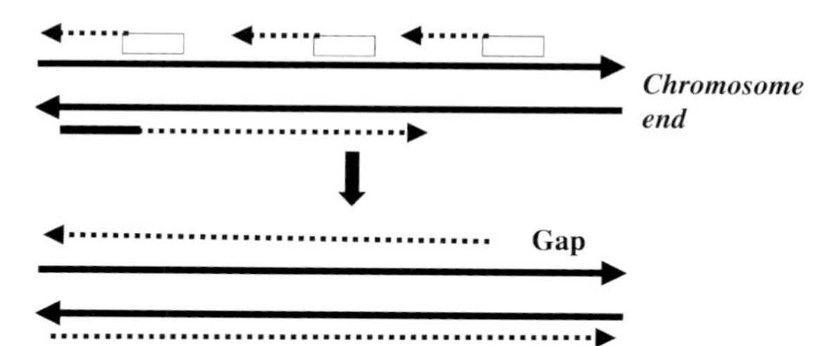

Fig. 1 Diagrammatic presentation of the "end replication problem". During lagging strand synthesis, at the end of chromosomes, gaps are generated in the absence of an alternative mechanism responsible for the maintenance of chromosome size. Thus, after each and every complete cycle of DNA replication the resulting, newly synthesized chromosome should be smaller than the parental one. This notion predicts the operation of an alternative mechanism for DNA replication that was later identified in telomerase. Open boxes indicate the RNA priming sites that correspond to the *Okazaki* fragments.

Understanding Carcinogenesis. Hippokratis Kiaris

ISBN 3-527-31486-5

Ends of Chromosomes Consist of Specialized Structures Termed "Telomeres"

The ends of the chromosomes of eukaryotic cells consist of some very specialized DNA–protein complex structures termed telomeres that protect cells from events causing DNA instability. These events include chromosome fusion generating dicentric chromosomes and loss of genetic information due to the progressive degradation of the chromosome ends. Telomeric DNA consists of short repetitive DNA that belongs to the microsatellite DNA and is usually rich in Gs. In humans and mice, telomeric DNA consists of concatamers of $(TTAGGG)_n$ repeats that are generated by the action of telomerase, a specialized RNA-dependent DNA polymerase, which is virtually a cellular reverse transcriptase (Fig. 2). Telomeres in humans usually reach a size of about 15–20 kbp at birth and, in the absence of telomerase activity or any other alternative mechanism contributing to the maintenance of telomere length, progressively decrease during each cycle of DNA replication. This reduction of telomeres during the life of an individual cell reflects the age of the cell in terms of its proliferative history and capacity. Furthermore, telomeres are thought to provide protection against malignant transformation since reduction of telomere size and subsequent growth arrest does not allow the accumulation of the multiple genetic hits that are needed for malignant conversion. Actually, as will becomes clearer below, telomerase exhibits a more complex role in carcinogenesis, being both negative (protective) and positive (permissive) depending on various factors such as the stage of transformation and malignant conversion, and the genetic background provided by other genetic lesions.

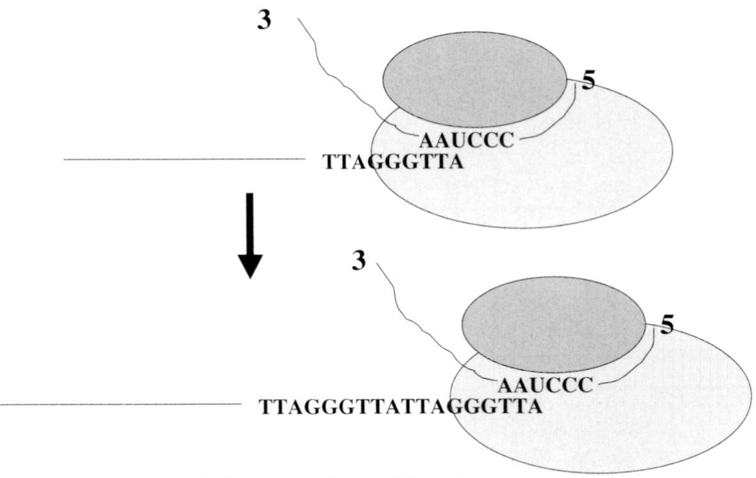

Fig. 2 Mechanism of telomere "replication" by telomerase.

Telomerase Structure and Activity

Telomerase is a ribonucleoprotein complex consisting of two major subunits – a catalytic subunit (TERT) that synthesizes the $(TTAGGG)_n$ repeats at the 3′-ends of the chromosomes and an RNA component (TERC or TR) that contributes the RNA template for the action of TERT. These two subunits interact closely with each other in a manner that is essential for the activity of telomerase. Additional subunits, such as the telomerase-associated protein, also contribute to the ribonucleoprotein complex. Interestingly, TERT, which encodes for the catalytic subunit of telomerase, contains a highly conserved domain that possesses the reverse transcriptase domain and other, also conserved domains with a function that is under investigation.

Considering that cancer cells must exhibit the capacity for an indefinite lifespan, it is anticipated that they must have the telomerase reactivated. Indeed, the development of an assay that assesses the levels of telomerase activity directly, termed the telomeric-repeat amplification assay (TRAP assay), confirmed this hypothesis (Fig. 3). It has been found that the great majority of primary human tumors and cancer cell lines, as well as cells of the gametic line, exhibit telomerase activity at considerable levels.

Thus, while telomerase should not be viewed as a transforming oncogene *per se*, it exhibits a positive action (oncogenic) during carcinogenesis by helping cells to bypass replicative senescence due to telomere shortening, thus having a permissive role in the process. In principle, this positive role of telomerase in carcinogenesis was also confirmed by experiments involving the development of mice deficient for TR; however, the complex interplay between telomerase activity and DNA repair renders the results of these experiments more complex to interpret than originally anticipated (see Chapters 5, 18).

While reactivation of telomerase should more likely be an early event in carcinogenesis, screening of primary tumors and early neoplastic lesions implied the opposite. Telomerase activity generally correlates with the progression of the disease, with advanced lesions exhibiting higher telomerase activity than their early-

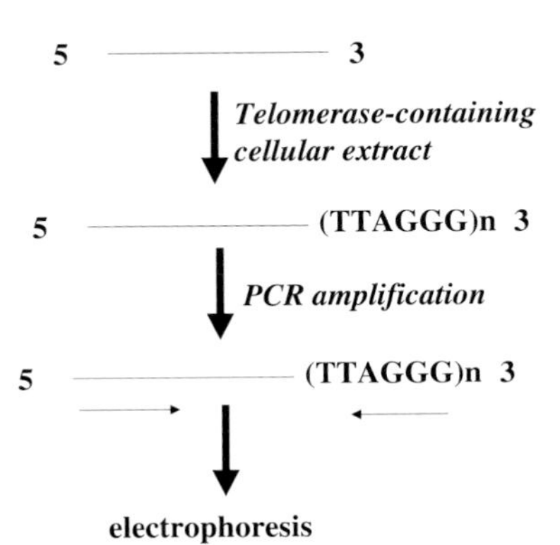

Fig. 3 A simplified overview of the TRAP assay that facilitates the rapid assessment of telomerase activity. Usually the oligonucleotides used for the TRAP assay bear certain modifications that increase the sensitivity and specificity of the assay (not shown).

stage counterparts. This observation probably reflects the occurrence of cellular crisis during early stages of the disease and is consistent with the high degree of chromosomal instability of primary tumors due to chromosome fusions.

Telomerase activity in a given tissue was found to correlate directly with the expression of TERT, whereas TR expression was expressed relatively abundantly. Thus, the limiting factor for telomerase activity is thought to be the activation of TERT transcription.

Regulation of Telomerase Activity

Various factors have been found to regulate telomerase activity. Of particular reference is the *myc* oncogene, which has been shown to induce the activity of telomerase during malignant transformation. The high oncogenic potency of *myc* in a wide range of tissues tested is thought to be due, at least in part, to its ability to stimulate telomerase activity and thus to induce cellular immortalization. Apart from *myc*, growth hormone-releasing hormone, the hypothalamic peptide that stimulates growth hormone production and release in the pituitary, was also suggested to stimulate telomerase activity. While a direct link between this observation and carcinogenesis has not been demonstrated, it is of particular significance because it provides a link between growth hormone and regulation of the cellular lifespan.

Telomerase is not regulated only at the level of activity. Intracellular localization of telomerase was also identified as a major determinant for its proper and regulated function. PinX1 is a suppressor of telomerase activity *in vitro* and *in vivo* because it forms stable complexes with TERT, preventing it from synthesizing telomeres because it lacks telomeric RNA (template). Furthermore, it sequesters telomerase in the nucleolus, while in the absence of PinX1 and during malignant transformation it predominantly localizes in the nucleoplasm. Thus, it has also been suggested that PinX1 functions as a TSG.

Therapeutic Implications of Telomerase

Modulation of telomerase activity appears to be a very promising target for cancer therapy. The almost general property of the cancer cells to exhibit reactivation of telomerase, in combination with its essential role in determining cellular mortality, implies the development of a generalized strategy based on telomerase to inhibit carcinogenesis. However, several concerns have been raised associated with the direct applicability of such a strategy. First, various normal cell populations, although in the minority, also express telomerase in the adult organism, importantly specific stem cells and activated lymphocytes. In the absence of a targeting mechanism it is conceivable that such cells will also be affected by the inhibition of telomerase activity and the consequences should be explored. Furthermore, from the time of initiation of antitelomerase therapy until telomeres reach a size that is critical for cell viability there is a lag period during which the progression of the disease cannot be affected. In addition, an intrinsic consequence of telomere attrition is the induction of genomic instability that may lead the tumors to acquire more aggressive properties. Finally, alternative mechanisms for maintenance of telomere size also exist that can potentially be activated during antitelomerase therapy and render tumors resistance to antitelomerase treatment.

It has been argued that all these obstacles are largely due to the fact that research in basic telomere biology has progressed more rapidly than the development of chemical or other modulators of telomerase activity and, thus, while the results of preclinical tests are very limited, various problems related to this type of treatment are anticipated. Therefore, it has been suggested that antitelomerase therapy will find applications in combination with other antitumor agents.

Alternative Mechanisms for Telomere Maintenance

As already mentioned above, while the vast majority of human tumors are positive for telomerase activity, certain malignancies have been found to be telomerase deficient, implying the operation of telomerase-independent mechanisms for the maintenance of telomere size. Indeed, the operation of such mechanisms has been proven and termed "alternative lengthening of telomeres" (ALT). The ALT mechanism involves homologous recombination between telomeres and results in the extension of telomere size in the absence of telomerase. The potency and the contribution of this mechanism to carcinogenesis, alone or in combination with telomerase activity, is under investigation.

Bibliography

Blackburn EH. Telomeres and telomerase: their mechanisms of action and the effects of altering their function. *FEBS Lett* **2005**, *579*, 859–862.

Lustig AL. Clues to catastrophic telomere loss in mammals from yeast telomere rapid deletion. *Nat Rev Genet* **2003**, *4*, 916–923.

Maser RS, DePinho RA. Connecting chromosomes, crisis and cancer. *Science* **2002**, *297*, 565–569.

Shay JW, Wright WE. Telomerase: a target for cancer therapeutics. *Cancer Cell* **2002**, *2*, 257–265.

9
Tumor Angiogenesis

Cancer cells, in general, are characterized by increased autonomy as compared to their normal counterparts; however, they still depend on their environment. This dependency is exemplified by their requirement to satisfy their metabolic needs and increases dramatically when tumors surpass a certain size (estimated to be about 1–2 mm^3). At this size, the tumors become highly vascularized, which is mandatory in order for them to further increase in size. Below that size, cancer cells form small nodules that are generally avascular since their size does not make it necessary for them to contain a dedicated vasculature. Their subsequent transition into this highly angiogenic state has been termed the "angiogenic switch" and is a delicately regulated process. During the angiogenic switch the cell responds to signals by producing and secreting several factors that result to the neoangiogenesis of tumors. Among the signals that trigger the angiogenic switch, hypoxia represents the one that is best characterized (Fig. 1).

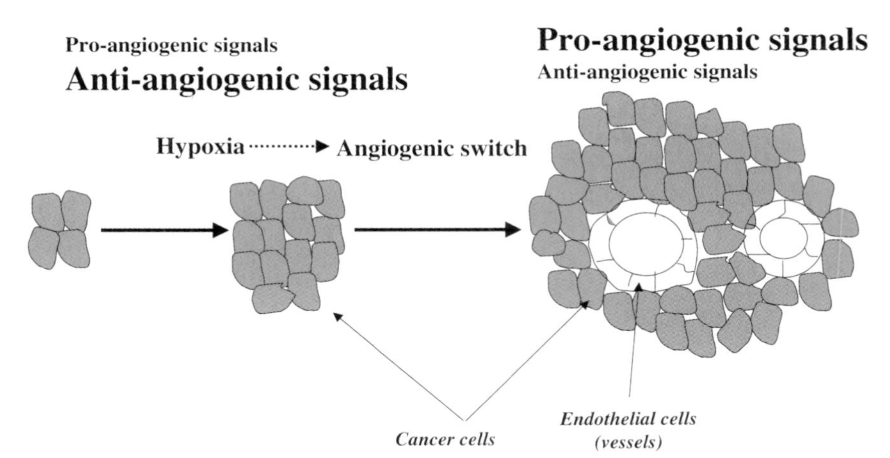

Fig. 1 Growing tumors become hypoxic – a state that triggers the "angiogenic switch". Following the completion of this "switch" the tumors become vascularized, which is necessary for their further growth.

Understanding Carcinogenesis. Hippokratis Kiaris

ISBN 3-527-31486-5

The complexity in the regulation of angiogenesis is due to the fact that it bears two aspects as reflected in the processes to which the endothelial cells (but also cancer cells, as it will become apparent below) are being subjected: a "proliferative" one for the outgrowth and extension of the tumor vessels, and a "morphogenic" one involving endothelial cell migration and formation of the specific vascular structures. These processes bear many similarities with neovascularization under physiological conditions, but are also characterized by certain differences.

The high dependency of tumors on continuous vascularization has been recognized for many years and it is postulated that it represents a potential target for anticancer drug development. Currently, tumor angiogenesis research is a very promising target for anticancer drug development.

Whilst in previous years the origin of tumor vessels was postulated to be exclusively from pre-existing vessels in the environment of the tumor that formed ingrowths within the tumor mass, today we know that more complex mechanisms are also operational, including the *de novo* formation of new blood vessels from angioblasts, as well as phenomena like vascular mimicry, etc., alone or in combination.

Distinct Organization of Tumor Vessels as Compared to those of Normal Tissues

Despite the fact that cancer cells employ mechanisms already operational in normal tissues and under physiological conditions for their vascularization, the resulting vasculature is different in many aspects (Tab. 1). In normal tissues, vessels are organized in a hierarchical manner such that all cells are close enough to vessels so that nutrients can be diffused and completely consumed until they reach their targets. However, in tumors, vessel organization is aberrant and noncanonical so that hypoxic conditions are frequently generated. Furthermore, the vessel diameters are uneven, and they form wide junctions at some regions and stacked layers of endothelial cells at others. In general, high heterogeneity exists in tumor blood vessels with regard to their permeability and blood flow, not only between a tumor and its metastases, but also within different regions of the tumor mass. The latter is considered a consequence of the imbalance between pro- and antian-

Tab. 1 Tumor vessels are different from those of normal tissue.

	Normal tissue	Tumor tissue
Vessel organization	canonical (all cells are adjacent to vessels)	aberrant (generation of hypoxic conditions)
Morphology	homogenous, even diameter of vessels	abnormal, uneven diameter of vessels
Expression profile	consistent expression of specific markers such as CD31 and certain adhesion molecules	inconsistent expression of specific markers

giogenic signals, such as vascular endothelial growth factors (VEGFs) that increase permeability, and others like Ang1 and Tsp1 that decrease it.

At the molecular level, endothelial cells are occasionally characterized by a different expression pattern from that of endothelial cells of normal vessels, as they frequently do not express common endothelial markers, such as CD31. Adhesion molecules are also expressed differentially in the tumor blood vessels. This is due to the opposite action of certain growth factors on the regulation of adhesion molecule expression. VEGF-A produces a positive effect on the expression of adhesion molecules, whereas others, such as transforming growth factor (TGF)-β, basic fibroblast growth factor (bFGF) and Ang1, inhibit it.

It has been proposed that modulation of the tumor's vasculature represents a challenging target for cancer therapeutics. Apart from the development of antiangiogenic therapy, which will be discussed below in more detail, "normalization" of tumor's vasculature may help in the more efficient delivery of anticancer drugs, increasing the efficacy of conventional targeted antitumor strategies.

Regulation of Angiogenesis: The "Angiogenic Switch"

As tumors grow in size their requirements for nutrients and, importantly, oxygen increase rapidly at levels that cannot be compensated for by the microvessels that are already operational in the microenvironment of the tumor. Thus, they need to induce the formation of new blood vessels dedicated to the support of the proliferating cancer cells. For the initiation of this complex process, the cancer cells need to sense that the formation of new blood vessels is required. Subsequently, as soon they receive this signal, they need to modulate their expression profile and stimulate angiogenic pathways, which results in the production of other signals that now act on the cells, predominantly endothelial, that will form these vessels. This whole process is illustrated in Fig. 1, and results in the transition of the equilibrium between angiogenic and antiangiogenic signals, towards the angiogenic state.

Several factors have been shown to regulate angiogenesis, among them the members of the VEGF family, FGF, platelet-derived growth factors (PDGFs), angiopoietins and others. A central regulatory role is attributed to hypoxia-inducible factors (HIFs).

Triggering Angiogenesis by Hypoxia

Hypoxia represents the major signal that induces angiogenesis and HIFs play a central role in this process. The HIF family contains several members (HIF-1, HIF-2, etc.) and in each one of them various subunits have been identified (HIF-1α, HIF-1β, etc.). HIF-1 is the prototype of this family of transcription factors. Among its two major subunits, HIF-1α is induced by hypoxia and HIF-1β is ubiquitously expressed. During conditions of normal oxygen levels (normoxia),

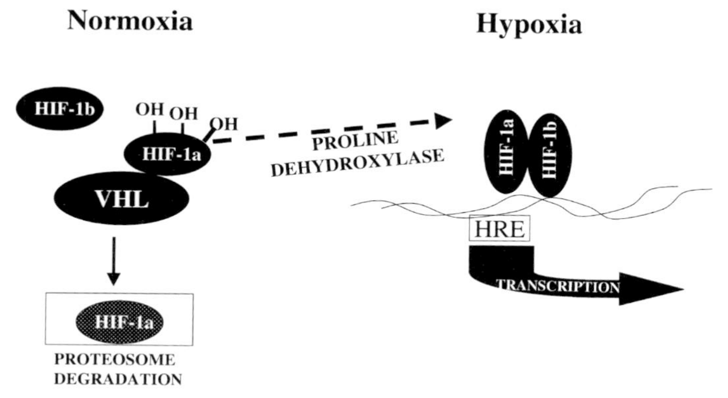

Fig. 2 HIF-1α plays a central role in the regulation of the angiogenic switch. During normoxia it is degraded after interaction with VHL. However, in hypoxia, following dehydroxylation it hetrodimerizes with HIF-1β and binds to the DNA where it operates as a transcription factor (see text for more details).

HIF-1α is bound to the product of the Von Hippel-Lindau (VHL) tumor-suppressor gene (TSG) participating in a bigger complex that results in its proteolysis (proteosome degradation). In this interaction, post-translational modification of HIF-1α plays a number of important roles, including hydroxylation by oxygen-dependent prolyl hydroxylases (PHD). Apparently, the tumor-suppressive function of the VHL is at least in part due to its ability to inhibit HIF function by stimulating its degradation.

During hypoxic conditions, HIF-1α dissociates from VHL and forms heterodimers with HIF-1β (or p53 depending on the conditions, as will discussed below). This dissociation of HIF-1α from VHL is favored by the reduced hydroxylation of PHDs which are highly dependent on oxygen levels. Subsequently, the HIF-1α/HIF-1β heterodimer is now free to bind to certain sequences in the regulatory regions of target genes and induce their transcription (Fig. 2). These regulatory regions have been termed hypoxia-regulatory elements (HRE) and have been identified in the promoters/enhancers of various genes that have been shown to play a direct role in angiogenesis, such as VEGFs, erythropoietin, etc. Interestingly, HIF-3α, another member of the HIF family, plays a negative (inhibitory) role in the regulation of angiogenesis.

Phosphorylation also plays an important role in the stimulation of HIF-1α activity. Notably, it has been suggested that the extent of HIF-1α phosphorylation is proportional to the severity of hypoxia and, in turn, differentially determines the affinity of HIF-1α binding to HIF-1β or p53. Therefore, during mild hypoxic conditions the pattern of HIF-1α phosphorylation favors binding with HIF-1β. Under strong hypoxia, however, the phosphorylation pattern of HIF-1α dictates binding to the tumor suppressor p53, which then induces apoptotic cell death (see Fig. 3).

It should be noted that regulation of HIF-1α activity is not only limited to the above-mentioned oxygen-dependent stabilization, but is also extended to the

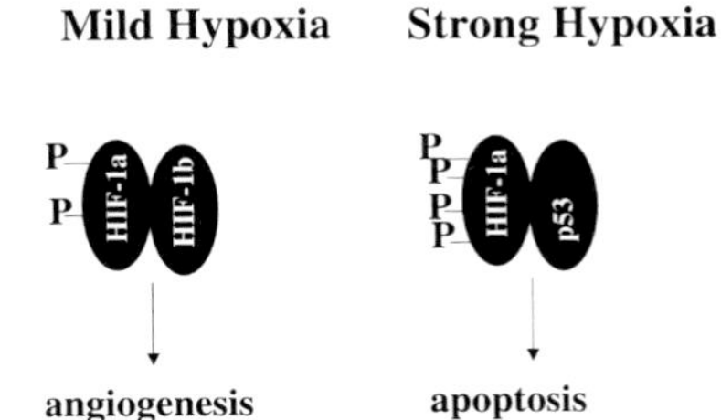

Fig. 3 It has been proposed that the extent of HIF-1α phosphorylation determines whether hypoxia will trigger angiogenesis or apoptosis. Under mild hypoxic conditions phosphorylation of HIF-1α favors binding to HIF-1β, which in turn induces angiogenesis. Alternatively, under heavy hypoxia, HIF-1α preferentially binds to p53 that in turn triggers apoptosis.

modulation of its activity by other factors including phosphtidylinositol-3-kinase, mitogen-activated kinase, etc.

Other Stimuli Triggering Angiogenesis

Although playing a key regulatory role, oxygen deprivation is not the only factor inducing angiogenesis. Other stimuli such as nutrient deprivation are also capable of triggering angiogenic pathways by mechanism(s) that occasionally overlap with those interpreting the hypoxic conditions. For example, glucose deprivation, a key nutrient in life, among its various consequences also results in the reduction of intracellular ATP levels which in turn causes a "hypoxia-like" type of stress. It can also cause oxidative stress that stimulates specific oncogenes that in turn induce the angiogenic switch by direct or indirect production and release of proangiogenic factors.

Oxidative stress, exemplified by the production of reactive oxygen species, can also trigger angiogenesis by stimulating the production of VEGF and other angiogenic factors.

Finally, several oncogenes, such as *ras* family members or the antiapoptotic *bcl-2*, induce angiogenesis by upregulating the expression of VEGF. Thus, oncogene activation not only affects cancer cell growth in an autonomous manner, but also triggers angiogenesis. Thrombospondin (TSP)-1 was shown to play an important regulatory role in oncogene-driven angiogenesis. TSP-1 is an extracellular matrix glycoprotein that has been identified as a naturally occurring inhibitor of angiogenesis. This has been supported by various experiments using transgenic and mutant mice, as well as tumor transplantation experiments in which genetic deficiency of TSP-1 in the host (mouse) inhibits the growth of unmanipulated cancer cells, resulting in the onset of slowly growing, poorly vascularized tumors. Thus, tumor growth inhibition is attributed to cells others than malignant cells. Various oncogenes have been shown to suppress TSP-1 expression, such as *ras*, c-*myc*, v-*src*, c-*jun* and Id1, whereas TSGs such as p53 and PTEN (phosphatase and tensin homolog deleted on chromosome 10) have been shown to activate TSP-1 expression.

Vascularization of Tumors

Following the angiogenic switch, tumors become vascularized by various mechanisms that involve either the *de novo* formation of new vessels or the acquisition and expansion of pre-existing vessels (angioblasts) (Fig. 4). We now describe in some detail these mechanisms and discuss how they are regulated.

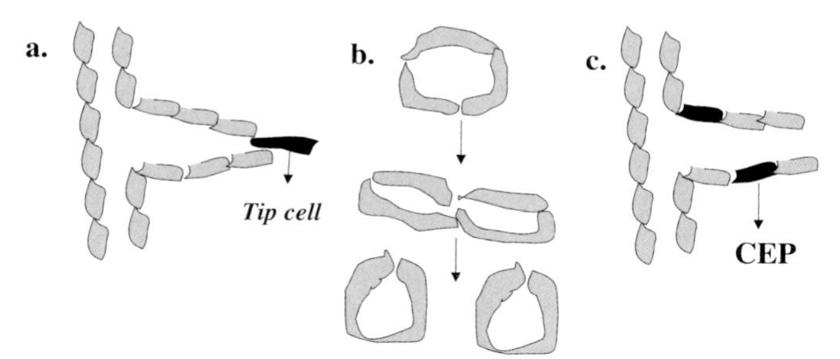

Fig. 4 Tumor neovascularization can be achieved by various different mechanisms. (a) "Sprouting" angiogenesis. (b) "Intussusceptive" angiogenesis. (c) Angioblast recruitment. In these examples the vessel wall consists of endothelial cells. In certain cases, however, tumor cells may also contribute to vessel formation (see text). Endothelial cells are shown in gray.

"Sprouting" Angiogenesis

This term refers to the predominant mechanism that accounts for the vascularization of growing tumors, and involves the proliferation and migration of endothelial cells from pre-existing vessels. VEGF plays a central role in this process. It has been shown that VEGF-A induces the proliferation of endothelial cells, while at the same time increasing the activity of metalloproteases and plasminogen activators that are responsible for the degradation of the extracellular matrix – a requirement for the appropriate migration of the endothelial cells. VEGF may also be responsible for the guidance of the migrating endothelial cells into the tumors by a mechanism according to which the tip cells at the front of the expanding vessels express VEGF receptor 2 and follow a VEGF gradient. The former mechanism has been shown to operate during the neovascularization of tissues such as the retina and hindbrain under physiological conditions. In malignancies, however, it is not yet clear as vascularization is relatively aberrant and may also involve additional mechanisms. Other factors that also regulate this process are FGFs and TGF-β.

"Intussusceptive" Angiogenesis

This is a variation of the mechanisms described above and involves the formation of a close contact between endothelial cells located across each other in a preexisting vessel. Subsequently, the vessels are separated. PDGF-B has been found to play an important role in this process.

Angioblast Recruitment

Circulating endothelial precursors (CEPs) can be recruited by the tumor, generating the tumor's vasculature. VEGFs and their receptors have been shown to play an important role in this mechanism. An interesting experiment showed that blockade of EGF receptor 2 action by an antibody diminished the recruitment of CEPs by the tumor. Furthermore, secreted factors, such as stromal cell-derived factor (SDF)-1, produced by the stromal fibroblasts also play important roles into the recruitment of CEPs into the tumor.

The extent to which this mechanism contributes to the vascularization of tumors depends on the type of the tumor. For example, CEP recruitment accounts for up to 90% of tumor vascularization in some lymphomas, whereas it is limited to only 5% in neuroblastomas. It also depends on the stage of tumor progression in a manner that it is more important in the earlier than the later stages of carcinogenesis. The latter has been shown by experiments indicating that at day 2 most of the tumor vessels in lymphomas have been derived by CEPs, while at day 14 only 50% have.

Cooption

This mechanism has been observed in some highly vascularized tissues, such as brain and lungs, in which tumor cells start growing surrounding the vessels that already exist in these tissues. Ang2 plays a central role in this mechanism and, following its production by the endothelial cells, results in induction of apoptosis, increase in the vessel diameter and decrease in their total number. The subsequent hypoxic conditions, which are due to the lack of vessels, stimulate VEGF-A-dependent angiogenesis and result in tumor vascularization in the periphery. The importance of this mechanism becomes apparent by the fact that it contradicts the notion that neoangiogenesis is required for tumor progression. In cases where vessel cooption is operational, neoangiogenesis in not needed.

Vasculogenic Mimicry

In some tumors, such as uveal melanomas, typical blood vessels are only localized in the periphery of the tumor and no vessels can be detected in its center. Notably, these tumors are not necrotic, thus implying the flow of blood. More detailed analyses suggest that channels interacting with peripheral blood vessels can be detected in these lesions that, although not composed of endothelial cells, partially stain with endothelial cell markers such as Von Willebrand factor, CD34 and Ulex lectins.

Whilst several lines of evidence support the notion that vascular mimicry does contribute to the vascularization of the tumor, certain considerations cast doubt regarding its functionality. For example, it has been argued that in the absence of a nonthrombogenic surface, such as constituted by endothelial cell vessels, thrombosis would have resulted in tumor necrosis, which apparently is not the case.

Mosaic Vessels

This mechanism refers to the presumed contribution of tumor cells in the tumor blood cell wall due to their localization. Endothelial cells in tumor vessels frequently undergo apoptosis, thus exposing cancer cells to the lumen, resulting in the generation of mosaic vessels consisting of both endothelial and cancer cells at varying ratios.

Antiangiogenic Cancer Therapy

That targeting tumor vessel formation possesses therapeutic implications was suggested many years ago, but the systematic development of drugs targeting neoangiogenic processes has only been initiated during the past decade.

Considering that the angiogenic switch occurs in lesions that exceed a certain size, then antiangiogenic therapy must be effective in inhibiting the progression of tumors and their metastases. Indeed, while expression of SV-40 early region, telomerase reverse transcriptase and *ras* oncogene was sufficient to cause malignant transformation of primary epithelial cells *in vitro*, *in vivo* it was not and needed either high levels of *ras* or VEGF expression. In both cases the angiogenic switch was induced, thus confirming that angiogenesis was needed for tumor progression.

Consistent with this notion, it has been shown that in experimental cancer models angiogenesis inhibitors can suppress tumor growth. For example, endostatin and angiostatin, which are endogenous antiangiogenic factors, reduce the formation of metastases in the *Lewis* mouse lung cancer model. Angiostatin binds to ATP synthase, angiomotin and Annexin II on the endothelial cells, and inhibits their proliferation and migration. Endostatin most likely targets

$\alpha_5\beta_1$ integrin, inhibits endothelial cell proliferation and migration, and induces apoptosis of proliferating cells without affecting wound healing.

In addition to those and other endogenous angiogenesis inhibitors such as canstatin, tumstatin, etc., in a different strategy an effort is made to prevent tumor angiogenesis by specific monoclonal antibodies. Of particular importance are antibodies against VEGF termed *bevacizumab* (*Avastin*) or *Vitaxin* (that targets $\alpha_v\beta_3$ integrin).

The major advantage of such therapy as compared to other conventional therapies that target cancer cells is that the endothelium composing the tumor vessels has a normal origin and genetic content, and thus does not exhibit the genomic instability that characterizes cancer cells. Therefore, antiangiogenic therapy should be devoid of phenomena such as the development of anticancer drug resistance, which is predominantly due to this instability of cancer cells, rendering it a universal therapy applicable in virtually all tumors. Unfortunately, however, this was not the case as evidenced by some experimental studies and the results of various preliminary clinical trials that showed very high variability to the sensitivity of several tumors in antiangiogenic therapy. It is conceivable that the operation of two major mechanisms may account for this discrepancy between the expected and the actual outcome of antiangiogenic therapy: (i) the selection of cancer cells that exhibit a lower demand for oxygen than the population of cells constituting the tumor prior to the initiation of the therapy and (ii) the employment of alternative pathways for tumor vascularization bypassing the targets of the antiangiogenic therapy. There is various experimental evidence to support either of these mechanisms. With regard to the first mechanism, it has been shown that p53 mutations, the most common genetic lesion in human primary tumors, apart from their direct effect on reducing the sensitivity of cancer cells to cytotoxic agents, also confer resistance to tumors in hypoxia. Thus, during antiangiogenic therapy it is unavoidable to select for hypoxia-resistant cells which are generally characterized by increased aggressiveness.

It has to be noted that currently it is considered more promising to use either a combination of antiangiogenic drugs or such drugs in combination with other anticancer agents.

In addition to the fact that antiangiogenic therapy is, in principle, less prone to the induction of resistance and less toxic than conventional therapeutic strategies, it has been proposed that it represents a good candidate for prophylactic therapy to patients with a high risk of developing cancer.

Another, more challenging antiangiogenic therapy would take advantage of recent progress in stem cell research. It has been demonstrated that genetic modification of endothelial progenitor cells to express a suicide construct converting 5-fluorocytosine to 5-fluorouracil and their administration into tumor-bearing mice results in their localization into their lung metastases where they produce a cytotoxic effect causing tumor growth inhibition. This approach has been paralleled with a therapeutic *Trojan horse* and represents a promising approach.

Bibliography

Auguste P, Lemiere S, Larrieu-Lahargue F, Bikfalvi A. Molecular mechanisms of tumor vascularization. *Crit Rev Oncol Hematol* **2005**, *54*, 53–61.

Bergers G, Benjamin LE. Tumorigenesis and the angiogenic switch. *Nat Rev Cancer* **2003**, *3*, 401–410.

Jain RK. Molecular regulation of vessel maturation. *Nature* Med **2003**, *9*, 685–693.

North S, Moenner M, Bikfalvi A. Recent developments in the regulation of the angiogenic switch by cellular stress factors in tumors. *Cancer Lett* **2005**, *218*, 1–14.

Volpert OV, Alani RM. Wiring the angiogenic switch: Ras, Myc, and Thrombospondin-1. *Cancer Cell* **2003**, *3*, 199–200.

10
Metastasis

Metastasis is the name given to the ability of cancer cells to spread from the primary site at which they have formed a (primary) tumor to distal sites in other organs. This can be done through the blood or the lymphatic vessels. Metastatic spread of cancer cells usually exhibits an organ-specific pattern. For example, breast and prostate cancers usually metastasize into the bones within a time period that can extend up to many decades in some instances. Metastases are the primary cause of mortality in cancer patients, and despite the recent advances in our understanding the basic biology of the disease and the development of various novel anticancer therapies, the prognosis of patients has been improved only slightly following the diagnosis of metastatic or invasive disease.

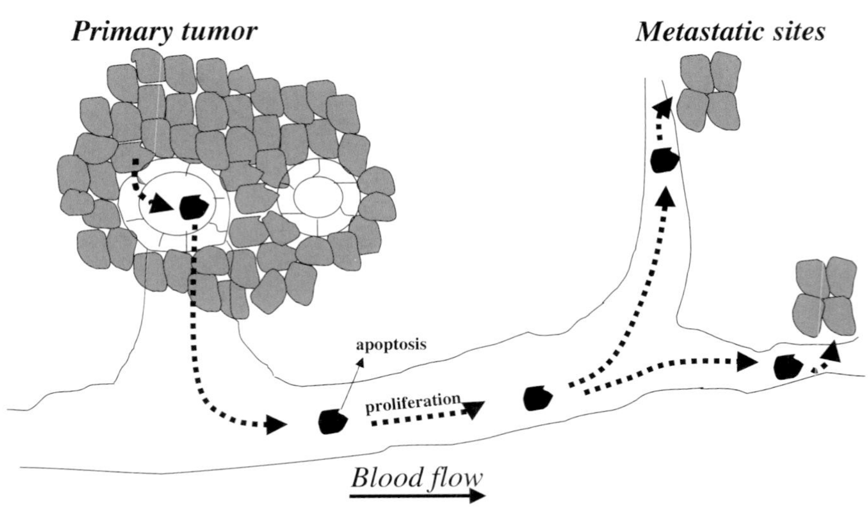

Fig. 1 Summary of the metastatic process. Some cells from the primary tumor escape and travel through the systemic circulation into the sites of metastasis where they continue to grow and establish new tumors. Throughout this process they are continuously subjected to signals that trigger apoptosis which they must overcome in order to metastasize efficiently.

Understanding Carcinogenesis. Hippokratis Kiaris

ISBN 3-527-31486-5

While different types of cancer represent distinct entities, genetically and biochemically, the basic processes underlining the metastatic process are similar, if not identical, in a variety of cancers. Such alterations usually involve growth factors, chemokines, cell adhesion molecules and matrix proteases that act in an orchestrated fashion to permit the escape of cancer cells and their "homing" to different tissues (Fig. 1). In many instances the metastasizing cells are also characterized by *dormancy*, which in some cases may extend to up to a few decades. During this period the cancer cells stay in a latent stage. Models based on mathematic simulation predict that during dormancy cells do not proliferate continuously, but only slowly, and they are subjected to repeated cycles of quiescence/proliferation.

The role of stroma of both the tissue of origin and the receiving tissue appears to play an important role in regulating metastasis efficiency. Only recently have components of the "metastatic pathway" begun to unravel, offering us important information regarding basic aspects of tumor biology and opening up new possibilities for the development of targeted anticancer strategies.

Metastatic Routes

The metastatic process, like carcinogenesis itself, is multistage and multifactorial. Therefore, acquisition of the malignant phenotype does not necessarily mean the ability to form metastases. The latter property of the cancer cells is only available upon the successful completion of a series of additional molecular alterations that render the cells capable of escaping their primary site of development, traveling (and surviving) through the lymphatic and/or the blood vessels, and finally localizing and growing in a different tissue that most likely provides a distinct microenvironment from that of the tissue of origin. Experimental evidence indicates that metastasis is a process generally characterized by low efficiency. While the escape of primary cancer cells into the circulation, lymphatic or blood is usually accomplished very efficiently, the growth of these cells into different tissues up to a point where they become clinically detectable, well-vascularized metastatic tumors is performed with low efficiency. This second phase represents the rate-limiting step of the metastatic process, and includes the steps of homing and metastatic colonization (Fig. 2). In general, two major obstacles must be surpassed in order for cancer cells to efficiently overcome this limiting step – they should be able to induce neoangiogenesis, so that micrometastases can grow beyond a certain size, and they should be also able to adapt into their new microenvironment and acquire responsiveness to the paracrine growth/survival signals that are present in this particular tissue. Most likely the latter, as will be discussed below, represents an important determinant for the apparent specificity that some primary tumors show for specific sites.

It has to be mentioned that subcutaneous cell injection experiments in nude, SCID or any other immunocomprimised/immunoincompetent mice virtually mimics this final stage of the metastatic disease where the cancer cells are forced to

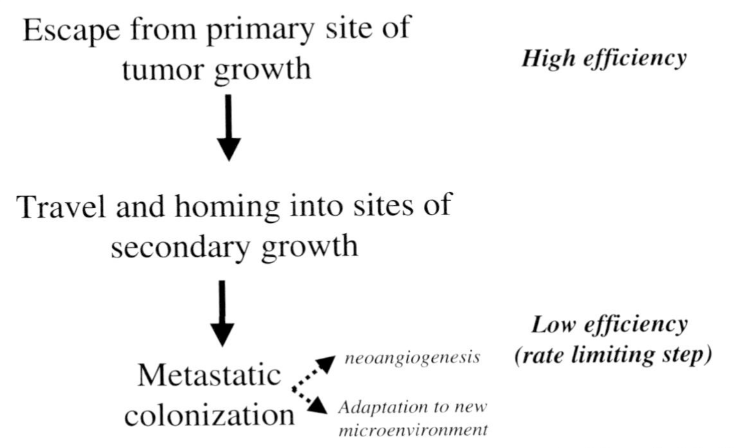

Fig. 2 In general, the first phase of metastasis, i.e. escape of the cancer cells from the primary site of tumor growth, is an efficient process. The second, i.e. homing and subsequent growth of the cells into the secondary site of tumor growth, is inefficient and represents the rate-limiting step of metastasis.

grow in a tissue microenvironment different from that they originated from. Therefore, it is frequently noted by investigators that various cancer cell lines can only grow orthotopically in the tissue that they have originated from, whereas subcutaneously they only grow very slowly or they do not grow at all.

What triggers the initiation of the metastatic process? It appears that the ability to metastasize is an intrinsic property of many malignant cells that has been acquired during the accumulation of mutations, while it is subsequently favored and selected as it offers advantages to the cells that bear it. However, experimental and clinical observations indicate that hypoxia in growing tumors also actively induces the metastatic cascade.

Usually, mechanical restrictions as soon as they enter the smaller capillaries arrest the cells in the distal organs, where the tissue microenvironment represents an important determinant for whether they are going to grow to new tumors. Along with mechanical forces that distribute the candidate metastasizing cells onto various tissues, the contribution of the microenvironment as an important determinant was noticed by Stephen Paget (1855–1926) during the late 19th century. In a classical article published in *The Lancet*, in which he dealt with the ability of certain tumors to give rise to secondary tumors in different sites, he stated: "When a plant goes to seed, its seeds are carried in all directions; but they can only live and grow if they fall on congenial soil". This comment sets the basis for the *"seed and soil"* theory that governs the metastatic growth of malignant tumors.

The physical route that the cancer cells follow during the course of metastasis greatly depends on the particular tissue the cells originate from (Fig. 3). In general, two major routes operate initially – the lymphatic system and the blood. Small blood vessels invading a malignant tumor can be penetrated by the cancer

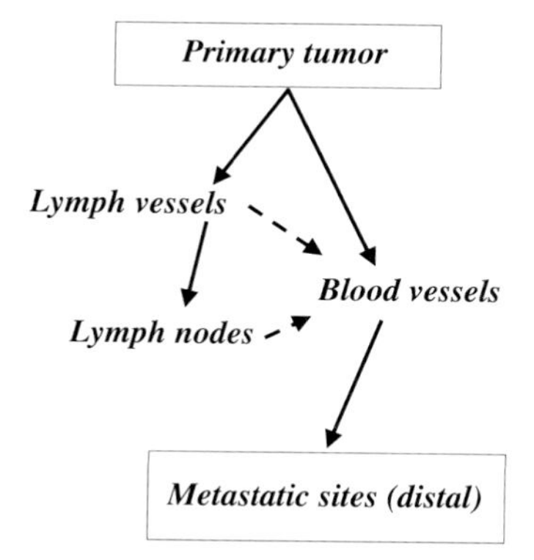

Fig. 3 Summary of the major metastatic routes. Escape from the primary tumors is achieved through the lymphatic and blood vessels that invade the tumor. Cancer cells travel through the blood circulation into the metastatic sites. Alternatively, lymphatic vessels that are localized around the primary tumor may carry the cancer cells that occasionally can grow into the lymph nodes, developing into lymph node metastases. Eventually, however, cells have to enter into the blood circulation to produce distal metastases. Exit from the lymphatic system can be achieved either through lymph nodes entering the blood vessels or by the blood vessels that support the lymph node metastases.

cells carrying them into the systemic circulation. An example can be offered by lung metastases of breast cancers. Alternatively, lymphatic vessels that are localized around the primary tumor may carry cancer cells that occasionally grow into the lymph nodes – a finding associated with poor prognosis in many cancers. Eventually, however, cells have to enter into the blood circulation in order to produce distal metastases. Exit from the lymphatic system can be achieved either through lymph nodes entering the blood vessels or by the blood vessels that support the lymph node metastases. It should be mentioned that many investigators consider that the entrance of cancer cells into the lymphatic vessels represents the predominant mechanism that initially operates to permit the spread of the tumors.

Lymphangiogenesis, i.e. the proliferation of new lymphatic vessels, is an active process that accompanies tumorigenesis, and is regulated at least in part by members of the vascular endothelial growth factor (VEGF) family of ligands and receptors. While angiogenesis is essential for the further growth of tumors, lymphangiogenesis is associated with increased metastatic potential. However, as the presence of peritumoral lymphatic vessels and their role in metastasis has been shown, the existence of intratumoral lymphatics is under debate.

Molecular Basis of Metastasis

Several categories of genes have been implicated in the regulation of the metastatic process. We will describe some of them on the basis of the stage at which they are thought to produce their major biological effects.

Cell–Cell Interactions

Central players are genes encoding for proteins mediating cell adhesion and cell–cell interactions. Integrins are membrane proteins that can dimerize to form complexes that mediate interactions with the extracellular matrix (ECM). Cadherins also regulate cell–cell adhesion. E-cadherin is a transmembrane protein. The extracellular portion of E-cadherin in neighboring cells interacts in adherence junctions, while the intracellular part interacts with catenins that, in turn, bind to the actin cytoskeleton. An inverse correlation between E-cadherin expression and a more aggressive phenotype in various tumors has been demonstrated, while reconstitution of E-cadherin adhesion complexes can restore it. N-cadherin, however, appears to play a different role as it enhances cell motility. Thus, it is possible that during cancer progression a "cadherin switch" operates from proadhesive cadherins such as E-cadherins to promigratory cadherins such as N-cadherin. B-catenin, which interacts with E-cadherin in the cell adhesion complex, also plays an important role in the metastatic process. Furthermore, it participates in *wnt*-mediated signal transduction, and plays a role in carcinogenesis in colon and other tissues. Finally, *ephrins* and *eph* receptors play important roles in metastasis as they are implicated in the regulation in cell communication, attachment, shape and mobility. Of particular interest in *eph* signaling is the fact that it operates bidirectionally in a manner that following the interaction between the transmembrane ligands and receptors, signals are transduced into both the ligand- and the receptor-expressing cells.

ECM Proteolysis

Remodeling of the microenvironment of the cancer cells, especially the proteolysis of the ECM, represents an important step during metastasis that is necessary for local invasion. Important proteins/enzymes that participate in this process are the matrix metalloproteinases (MMPs), serine proteases, adamalysin-related membrane proteases (ADAMs), bone morphogenetic protein (BMP)-1-type metalloproteinases, heparinase and cathepsin.

MMPs belong to a heterogeneous class of about 25 zinc-dependent proteases. An interesting aspect of their function is that many of them are produced by stromal cells upon appropriate stimulation by secreted factors from cancer cells. The regulation of their activity is a complex process that reflects the balance between positive and negative signals. Upon activation, MMPs degrade the ECM, and permit the local invasion and subsequent metastasis of cancer cells. Furthermore, through proteolytic cleavage, they may activate latent forms of growth factors and their receptors. For example, it has been demonstrated that thrombin receptor protease-activated receptor (PAR)-1, undergoes cleavage and activation in the cell membrane by MMP-1, which is derived by the stromal cells. PARs belong to a family of four G-protein-coupled receptors that carry their own ligand that is masked at the N-terminus. Upon activation by proteolytic cleavage, PARs activate G-proteins that, among their various effects, initiate the migratory and invasive program of cancer cells. The contribution of PARs to the metastatic ability of cancer cells has been demonstrated by experiments showing that suppression of PAR expression renders cells noninvasive. Chemokines, cytokines and other factors exert a positive action on MMP activity, whereas tissue inhibitors of metalloproteinases (TIMPs) have a negative regulatory role.

ADAMs are transmembrane proteins that are involved in both cell adhesion and ECM lysis, as they contain both disintegrin and metalloproteinase domains. Many ADAMs belong to another class of proteins, called *sheddases* because they can cleave cytokines that are anchored in the cell membrane, adhesion molecules, growth factors, etc., that are subsequently shed from the plasma membrane. An example is offered by ADAM 19 – a sheddase that cleaves (releases) a soluble form of Delta, which is the transmembrane ligand of Notch receptors.

Serine proteases include the urokinase plasminogen activator (uPA), thrombin and plasmin. uPA is activated upon binding to its receptor (uPAR) and, in turn, activates plasminogen by converting it into plasmin (another serine protease). Plasmin possesses broad specificity in degrading various ECM-related proteins and activating several growth factors by proteolytic cleavage. The importance of the uPA system in carcinogenesis is supported by the fact that several of its components are upregulated in various cancers and this is associated with poor prognosis, whereas its inhibition in experimental tumor models suppresses the metastatic and invasive ability of tumors.

Heparan sulfate proteoglycans play an important role in the organization of ECM, and the sequestration and stabilization of bioactive molecules. Heparanase cleaves the heparan sulfate chains, affecting the integrity and functionality of tissues. Increased heparanase activity has been detected in various primary cancers and metastatic cell lines, whereas its activation in cells exhibiting low metastatic ability increases it.

Cysteine cathepsins, a family of proteases with divergent specificity and expression profiles, also play important roles in the degradation of the ECM, and the local invasion and initiation of metastasis. In primary tumors some cathepsins, such as cathepsin B, can be detected in the invasive front, implying a role in the degradation of the basement matrix. Components of the ECM that are targeted by the cathepsins are laminin, fibronectin and collagen, and participation in regulatory networks has also been shown.

Metastasis-suppressor Genes

"Metastasis suppressor genes" are genes that are thought to play an important role in the regulation of the metastatic process. Their identification is based on the assumption that these genes should be downregulated (suppressed) in metastatic tumors as compared to tumorigenic, but nonmetastatic tumor cells, essentially in model systems. These genes include MKK4 that encodes for a mitogen-activated protein kinase (MAPK) kinase, KiSS1, encoding for a ligand of a G-protein-coupled receptor and others. Such genes are able to suppress the metastatic ability of cancer cells without affecting their tumorigenicity. However, several investigators have some doubts regarding how specific they are in regulating the metastatic process. Analogous to the suppressors of metastasis, inducers of this process have also started to be identified, exemplified by *Twist*, a gene that also plays a role in cell differentiation. In addition, genes with profound implications in malignant transformation, such as oncogenes as exemplified by activated *ras* family members, can affect the metastatic ability of cancer cells. Whether this is just a reflection of the fact that *ras* signaling activation favors proliferation and, thus, the survival of micrometastases remains to be seen.

Migration and Motility of the Cancer Cells

Upon modification of the ECM and manipulation of their microenvironment, cancer cells must be able to acquire properties that permit their movement. This is achieved by various mechanisms and involves several regulatory networks. *Collective movement* of tumor cells represents a relatively common manner by which they can move and which is β_1 integrin dependent. *"Amoeboid"* movement that is independent of β_1 integrin has also been documented and is applicable in single cells that detach from the tumor. Proteins that belong to the ECM can be subjected to proteolytic activation by enzymes produced by the cancer cells inducing them to migrate. Such proteins include fibronectin, laminin, thrombospondin, etc. *Chemotaxis* and *aptotaxis*, corresponding to movement towards a chemical gradient or a bound substrate, represent the major driving forces that dictate the specific movement of cancer cells during local invasion. Stromal-derived factors are also involved in the motility of the cancer cells such as insulin-like growth factor (IGF)-I and -II and hepatocyte growth factor (HGF), which are also implicated in the paracrine stimulation of the proliferation of cancer cells.

Homing

The fact that specific primary tumors exhibit a tendency to metastasize to particular secondary sites implies that the location of the metastatic growth is not a random process, but it is governed by specific characteristics. Three main mechanisms have been suggested to play a role in the selection of the site of metastasis (Fig. 4): (i) selective growth of the cancer cells in the tissue that favors their proliferation, (ii) selective adhesion onto the endothelium adjacent to the tissue of metastatic growth and (iii) attraction of the cancer cells by factors produced by this tissue. Experimental evidence is available to support all three mechanisms.

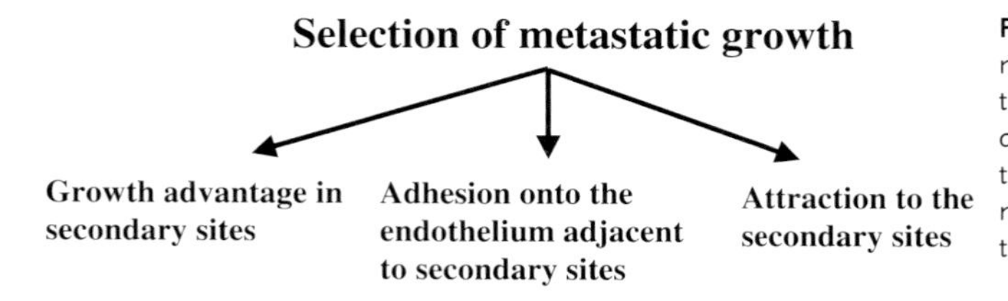

Fig. 4 The three major mechanisms that are thought to determine the selection of the site of metastatic growth of tumors.

Clinical Implications

As with all advances in the basic biology of carcinogenesis, understanding the molecular details of the metastatic process can be helpful at two different levels – setting the basis for the development of cancer therapeutics targeting the metastatic ability of tumors and the identification of specific markers that will permit the prediction of the prognosis of the disease. Considering that metastases represent the major cause of death of cancer patients, both targets acquire increased importance. Considering the fact that, as already mentioned above, regardless of the exact type of each tumor, common molecular pathways operate to induce the metastatic ability, both therapeutic and prognostic achievements may imply wide applicability. Regarding specific therapeutic strategies based on the inhibition of metastasis, suppression of the neoangiogenesis that is required by the metastatic cells is a promising approach and is discussed in more detail in Chapter 9. More specific inhibitors of metastasis, like the MMP antagonists, have been tested in clinical trials; however, the initial results are not very promising.

Bibliography

Bogenrieder T, Herlyn M. Axis of evil: molecular mechanisms of cancer metastasis. *Oncogene* **2003**, *22*, 6524–6536.

Chambers AF, Groom AC, MacDonald IC. Dissemination and growth of cancer cells in metastatic sites. *Nat Rev Cancer* **2002**, *2*, 563–572.

Naora H, Montell DJ. Ovarian cancer metastasis: integrating insights from disparate model organisms. *Nat Rev Cancer* **2005**, *5*, 355–366.

Pei D. Matrix metalloproteinases target protease-activated receptors on the tumor cell surface. *Cancer Cell* **2005**, *5*, 207–208.

Stacker SA, Achen MG, Jussila L, Baldwin ME, Alitalo K. Lymphangiogenesis and cancer metastasis. *Nat Rev Cancer* **2002**, *2*, 573–585.

Steeg PS. Metastasis suppressors alter the signal transduction of cancer cells. *Nat Rev Cancer* **2003**, *3*, 55–63.

Turk V, Kas J, Turk B. Cysteine cathepsins (proteases): on the main stage of cancer? *Cancer Cell* **2004**, *4*, 409–410.

Part III
Specific Topics

Understanding Carcinogenesis. Hippokratis Kiaris

ISBN 3-527-31486-5

11 Tissue Context as a Determinant of the Tumor-suppressive or Oncogenic Function of Certain Genes

Despite the pleiotropy in the effects of virtually all genes implicated in carcinogenesis, their involvement in the development of the disease is usually viewed simply as positive or negative. For example, activated *ras* genes encode for potent oncoproteins that *in vitro* can cause malignant transformation, whereas at the same time they are able to induce angiogenesis. Both of these processes have a positive effect on the development of cancer. Furthermore, the p53 tumor-suppressor gene (TSG) can induce either cell cycle arrest or apoptosis, both of which inhibit carcinogenesis. Several similar examples can be described for many oncogenes or TSGs.

This oversimplified mode of action has been recently questioned by experimental results clearly showing that the same type of alteration in the same gene(s), depending on the type of cells used for the analysis, may have either positive or negatives effect in carcinogenesis, behaving either as an oncogene or as a TSG, respectively. While not many examples of genes behaving in this manner are known, it is worth mentioning them because their action is related to an important aspect of tumor biology – the view that carcinogenesis is a disease of abnormal differentiation. Indeed, cancer represents a disease that is characterized by the acquisition of alternative cell fates by the cells. This alternative cell fate can be either de-differentiation or trans-differentiation. Thus, it is conceivable that under certain conditions genes implicated in the regulation of cellular fate are able to induce or suppress carcinogenesis. Two families of genes that are well-known regulators of cellular differentiation and also found to possess oncogenic or oncosuppressive potential are the RUNX transcription factors and the Notch receptors. Their involvement in carcinogenesis will be discussed briefly below.

RUNX Genes

RUNX genes encode for proteins that along with core binding factor (CBF)-β, form complexes that constitute transcription factors with essential roles in the induction or suppression of the transcription of genes involved in cellular differentiation, as well as growth and survival. The RUNX transcription factor family con-

Understanding Carcinogenesis. Hippokratis Kiaris

ISBN 3-527-31486-5

sists of three members known as CBF-α, polyoma enhancer-binding protein (PEBP)-2α and acute myeloid leukemia (AML) protein. All three possess a conserved 128-amino-acid Runt domain that is important for their function as it is sufficient for binding to DNA and to CBF-β. Binding to DNA is via the consensus sequence R/TACCRCA and the activity is regulated by physical interactions with other proteins including a series of corepressors that are essential for the regulation of RUNX activity.

The implication of the RUNX genes in cancer was initiated by the discovery of specific translocations in acute leukemia that result in the generation of a chimeric protein that contains the N-terminal region of AML1, including the Runt domain, fused to a protein termed ETO. Since then several different RUNX1-containing fusion proteins have been identified in leukemias. Full-length RUNX1-containing fusions have also been identified, such as the TEL–RUNX1 fusion, with Tel being a member of the Ets transcription factors. Although the corresponding mechanisms of action of the resulting RUNX1 fusion protein are characterized by considerable variability, in general it is thought that these fusion proteins antagonize the normal RUNX1, thus exhibiting a negative effect on RUNX1 action. For example, the TEL–RUNX1 fusion protein operates as a constitutive repressor of the RUNX1 target genes, RUNX1–ETO recruits corepressors and the C-terminal truncation of RUNX1 abolishes the transactivation potential of RUNX1.

In addition to translocations, point mutations in RUNX1 have also been detected quite frequently in some leukemias, especially AMLs. These mutations are localized predominantly in the Runt domain of RUNX1 – affecting DNA binding and producing truncated forms of the protein due to premature termination.

The fact that the function of RUNX1 fusion proteins is generally inhibitory, thus antagonizing the normal RUNX1, in association with the observation of the frequently detected truncations of the protein due to point mutations imply that RUNX1 is a TSG. This notion is further supported by the detection of germline mutations in the RUNX1 gene in patients suffering from familiar platelet disorder – a condition that predisposes to AML.

The-tumor-suppressive activity of RUNX1 can be extended to RUNX3 since it was found to be frequently downregulated in primary human cancers of epithelial origin. Interestingly, the mechanism of RUNX3 suppression in many instances was found to be epigenetic, i.e. due to hypermethylation.

Whilst the role of RUNX1 in carcinogenesis appears to be simple, operating as a TSG, the picture is complicated by additional experimental results suggesting that the oncogenic phenotype caused by simple loss of RUNX1 function is not similar to that of some other fusion proteins, such as RUNX1-ETO. The latter appears to act as a dominant oncogene able to cause malignant conversion and neoplastic transformation in fibroblasts. The oncogenic potential of RUNX family members is also supported by the observation that all three represent common targets of retrovirus-mediated insertional activation in models of mouse lymphoma. Furthermore, amplification of RUNX1 has been reported in some aggressive forms of child B cell acute lymphoblastic leukemia (ALL). RUNX2 and 3 have not been studied in the same detail as RUNX1; however, experimental evidence is available to support their oncogenic function.

Collectively, RUNX genes provide an example of genes that function in a manner consistent with the idea that the same gene may behave as a TSG or as an oncogene under different conditions. This notion suggests that what is defined as tumor-suppressive or oncogenic activity, respectively, is not the linear outcome of a specific alteration in a given gene, but rather the combined output of this alteration in the specific developmental or histological context. These findings have important implications in the management of cancer considering that the same drugs (or approaches, in general) may produce opposite effects when the cellular context changes.

Notch Genes

A clearer example of how the developmental context may affect the oncogenic/oncosuppressive potential of specific loci is provided by the Notch receptors.

Notch genes obtained their name from a specific mutation in *Drosophila melanogaster* (fruit fly) that results in the generation of notches in their wings. Genetic and molecular analyses suggested that this mutation is due to the hemizygous loss of function (haploinsufficiency) of a gene that encodes for a large, single-pass transmembrane receptor, named the Notch receptor. In mammals, four homologs of the *Notch* receptor have been identified that exhibit overlapping functions and a restricted pattern of expression in different tissues. The major function Notch receptors is the determination of cell fate decisions; however, they are involved in virtually every aspect of human physiology, as very few processes (if there are really any) have been shown to operate in a Notch-independent manner.

At the molecular level, according to the prominent model of Notch signaling, activation occurs following binding of the Notch ligands to the Notch receptor (Fig. 1). There are at least five ligands of the Notch receptor in mammals, and are these designated delta-like (DLL) 1, DLL3 and DLL4, and Jagged (JAG) 1 and JAG2. Upon ligand binding to the Notch receptor, the latter, which is also a transmembrane protein, undergoes a series of precisely regulated proteolytic cleavages; finally, the intracellular portion of the receptor is released from the cell membrane and translocates into the nucleus. This process is precisely regulated through interactions of the intracellular Notch receptor with various cytoplasmic and nuclear proteins that act as modifiers of its action. Such modifiers are the mammalian homolog of the *Drosophila Deltex* and *Mastermind*. Importantly, *Deltex*, which operates as a positive regulator of Notch signaling in *Drosophila*, functions as a negative regulator of Notch signaling in mammals. Finally, activated Notch receptor acts by participating in a protein complex that contains the human homolog of the *suppressor of hairless* (*SuH*) fly gene that is a transcription factor termed CBF-1. CBF-1 is transformed from a transcriptional repressor to a transcriptional activator upon binding to Notch. There, it activates the tran-

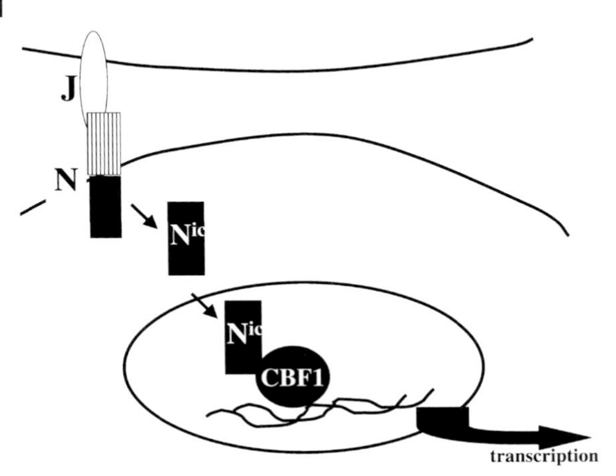

Fig. 1 Notch signaling. Upon activation by one of its ligands such as Jagged (J), the Notch receptor is cleaved and the intracellular portion of the receptor ($N1^{ic}$) translocates into the nucleus. There, it binds to transcription factors such as CBF-1, transforming it from a transcriptional repressor to a transcriptional activator. Several proteins, not shown in this diagram, participate in Notch receptor complexes and regulate its activity.

scription of various target genes, producing divergent biological effects that are characterized by high pleiotropy and context specificity. This specificity is very high, to such a degree that it is not possible to predict the consequences of Notch signaling activation in different tissues because Notch's actual function is thought not to induce, but rather to permit certain changes. Instead, we may classify these responses in some general categories. Thus, Notch activation may be responsible for the maintenance of cells in an undifferentiated state, maintaining the stem cell population of tissues, or may be involved in binary cell fate decisions. According to the latter mode of action, cells of the same type may be differentiated due to the status of expression of Notch receptor and ligand. In other words, a cell that expresses the ligand is destined to acquire a different cell fate from that of a cell expressing the receptor. In a slightly different mode of action the cells expressing and receiving Notch may be of a different type, with the first type inducing certain developmental changes in the population of cells expressing the receptor. Among the specific molecular targets of Notch signaling activation, various genes have been demonstrated to respond to Notch signaling activation. Among these genes we mention the members of the *Hes* (Hairy-enhancer of split) transcription factors that affect cellular differentiation and cell cycle regulators cyclin D1, which is an inducer of cell cycle progression, and $p21^{Waf1}$ cyclin-dependent kinase inhibitor (CDKI), which is a suppressor of cell cycle progression. In both cases we emphasize the fact that Notch-dependent induction of the expression of these genes is direct, at the level of their promoter.

It has to be mentioned that exactly this part of Notch signaling, that involves the translocation to the nucleus, is only proven and widely acceptable for mammals, whereas the inability of investigators to detect Notch receptor in the nucleus in *Drosophila* casts some doubt regarding how general this mechanism might be. However, the counter-argument is related to the suggestion that only very limited amounts of Notch receptor in the nucleus, below the detection limits of the current imaging methodologies, are sufficient to elicit Notch signaling activation. Further support of how sensitive the cell is to precisely regulated amounts of

Notch is provided by the fact that the Notch locus is among the very few loci that produce distinct phenotypes that are extremely dependent on gene dosage, as evidenced by the fact that both triplications as well as deletion of single alleles producing hemizygosity produce mutant animals with distinct phenotypes.

The implication of Notch signaling in tumorigenesis was initiated about 15 years ago after the observation that a subset of T cell ALLs contains a translocation between chromosomes 7 and 9 that generate a fusion protein that was shown to correspond to a truncated Notch1 gene, resulting in the expression of only the intracellular portion of the corresponding receptor. We note that this intracellular portion of the Notch receptors is the part of the protein that upon ligand binding is cleaved and translocates into the nucleus. Thus, this fusion generates a form of the receptor that mimics the activated receptor, in the absence of ligand binding, and therefore constitutes ligand-dependent activation. Indeed, that this part of the protein, the intracellular portion of the Notch receptor (Notch-IC), functions as such was later shown using biochemical assays that demonstrated that it encodes for a ligand-independent, constitutively active form of the receptor that is etiologically involved with the onset of T cell ALL.

Further evidence for this relationship is provided by the observation that integration of the Moloney murine leukemia virus (MuLV) frequently occurs within the Notch1 locus, generating truncations that mimic the activated form of the receptor, producing the cytoplasmic portion of the protein.

A similar mechanism originally associated Notch signaling with solid tumors and, especially, mammary cancers. Insertional mutagenesis experiments using the mouse mammary tumor virus (MMTV) showed that *Notch4* receptor, originally designated as *Int3* (integration 3) at that time, was frequently disrupted by MMTV, generating gain-of-function alleles encoding for truncated forms of the Notch receptors similar to the ones found in T-ALL. Subsequent experiments using transgenic animals expressing in the mammary epithelium the cytoplasmic portion of the Notch4 initially, and Notch1 later, receptors confirmed the etiological association between these mutant Notch receptors and mammary carcinogenesis. Studies on the pathophysiology of Notch-induced mammary cancers suggested that constitutive activation of Notch signaling arrests cells into an undifferentiated state, producing animals that fail to undergo proper terminal differentiation in the mammary epithelium upon hormonal stimulation. It is notable that the Notch1-related mouse model exhibits an interesting association between the onset of benign lesions and lactation, exemplifying the developmental context-dependent action of Notch receptors. Furthermore, some preliminary findings on primary human tumors also show that breast and probably other epithelial cancers overexpress Notch signaling components, including the Notch receptors and ligands.

The primary data obtained from experiments involving animals clearly show that activated Notch receptors, at least in certain tissues such as the mammary epithelium and the lymphatic system, possess oncogenic potential. However, the precise molecular mechanism that operates in the manifestation of this phenotype remains obscure. *In vitro* experiments in cell culture show that activated

Notch receptors alone exhibit a limited oncogenic potential; however, they strongly synergize with other oncoproteins to induce malignant transformation. The oncogenic targets of Notch receptors are unknown. Relatively recent evidence showed that cyclin D1, an activator of the G_1/S transition, is a direct target of Notch signaling activation. Furthermore, crosstalk between Notch and *ras* signaling has been demonstrated in flies and in mammals under neoplastic conditions, whereas suppression of the pathway can inhibit the oncogenesis induced by the activated *ras* oncogene.

The generalization of a positive action of Notch in epithelial tissues has been questioned by recent experiments that showed that activation of the pathway in the keratinocytes induces their differentiation, while its tissue-specific ablation, due to allelic deletions, results in the development of hyperproliferative lesions and increased sensitivity to carcinogenesis induced by chemicals. These effects were causatively associated with the specific induction of the $p21^{Waf1}$ CDKI by Notch signaling activation. The abovementioned findings imply a tumor-suppressive activity for Notch in the epidermis that opposes its oncogenic activity in mammary epithelial cells.

Considering that in both cases activation of Notch is able to induce the expression of both tumor promoters (i.e. cyclin D1) and TSG (i.e. $p21^{Waf1}$) by direct transcriptional activation, which rules out the possibility that the responses obtained represent self-protective mechanisms of the cell, it remains unclear how the outcome of the response is going to be interpreted by the cell. A simple explanation is that the latter is being determined by the relative balance of the specific coactivators/corepressors available in each particular cell type or, alternatively, by the sensitivity of each cell type to the specific response. The precise mechanism remains to be revealed as well as the specific implications of this context-dependency for cancer therapeutics.

Bibliography

Artavanis-Tsakonas S, Rand MD, Lake RJ. Notch signaling: cell fate control and signal integration in development. *Science* **1999**, *284*, 770–776.

Blyth K, Cameron ER, Neil JC. The RUNX genes: gain or loss of function in cancer. *Nat Rev Cancer* **2005**, *5*, 376–387.

Kiaris H, Politi K, Grimm LM, Szabolch M, Fisher P, Efstratiadis A, Artavanis-Tsakonas S. Modulation of Notch signaling elicits signature tumors and inhibits Hras1-induced oncogenesis in the mouse mammary epithelium. *Am J Pathol* **2004**, *165*, 695–705.

Radtke F, Raj K. The role of Notch in tumorigenesis: oncogene or tumor suppressor? *Nat Rev Cancer* **2003**, *3*, 756–767.

12
Cancer Stem Cells

As discussed extensively elsewhere, malignant tumors are characterized by high genetic, cellular and morphological heterogeneity. When we discussed the role of tumor stroma and tumor angiogenesis in cancer development (Chapters 7 and 9), we made the distinction of tumor cells constituting the neoplastic component (transformed cells) from those composing the nonneoplastic component (fibroblasts, endothelial cells, etc.) of each tumor. However, even considering the neoplastic cells of a tumor as a homogenous population of cells that differs only because some of them accumulate mutations that may offer general proliferative or survival advantage represents an oversimplification.

Early studies have demonstrated that depending on the particular cells used, in order to perform tumorigenicity-related experiments in mice, it was always necessary to inoculate a minimum number of cells into the appropriate hosts in order to grow tumors. If indeed tumors consisted of homogenous populations of cancer cells or, alternatively, if all cancer cells within a tumor were identical, then theoretically even a single cell should be sufficient to grow a tumor in mice. In that case the cell number could only determine the latency of the onset of palpable tumors – the more cells injected, the faster the tumor would develop. However, experience indicated otherwise. Only injection of at least a few hundred thousand cells is usually sufficient to grow tumors in mice. Similar observations have also been made *in vitro* with clonogenic experiments in which the lifespan of individual cells, contrary to that of a whole culture, was examined. Thus, cancer cells are not homogenous and only a small fraction obtained by a single tumor can grow *in vivo* and generate a solid tumor. Consistent with the aforementioned observation and in line with the current notion that views the tumor as an entity of different cell types, only a few cells will have the capacity to grow, generating the epithelial (cancer) cells that will constitute the tumor. Those cells with the capacity for self-renewal represent the cancer stem cells, and are responsible for the generation of both the stem cells and the "differentiated" cancer cells (Fig. 1). The latter are the cells that, while considerably contributing to the tumor mass, do not offer tumorigenic properties as exemplified by the ability to grow in mice. Such heterogeneity is apparent in some teratocarcinomas in which only a small portion of the cancer cells express specific markers, e.g. β-human chorionic gonadotropin (β-hCG), whereas others can produce teeth and hair.

Understanding Carcinogenesis. Hippokratis Kiaris

ISBN 3-527-31486-5

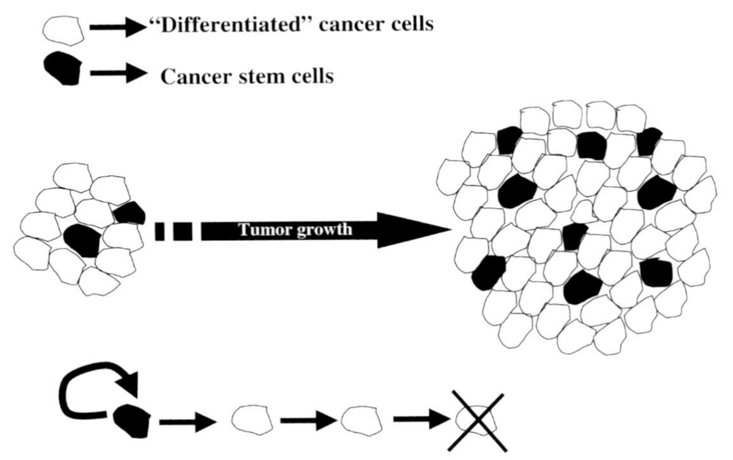

Fig. 1 Only cancer stem cells are able to self-renew, producing cancer stem cells, and to differentiate, producing the "differentiated" cancer cells. The latter eventually die after some limited proliferation (self-renewal).

While both cancer stem cells and the "differentiated" cancer cells have the capacity to proliferate, in the latter this capacity is limited and only the stem cells possess this ability indefinitely (Fig. 1). Indeed, they are capable of self-renewal as characterized by the fact that the resulting populations of cells are identical. Whether cancer originates from the normal stem cells that have the capacity for self-renewal or from committed progenitor cells remains to be established and supported by some definite experimental proof. However, it is conceivable that in the second possibility, in addition to the stimulation of pathways that involve perturbation of cell death/proliferation equilibrium, the activation of self-renewal programs is also needed.

In order to prove this hypothesis one needs to separate cells from a given tumor into subpopulations of identical cells and assess their tumorigenicity (Fig. 2). Indeed, such studies have recently been performed in breast tumors and have confirmed this hypothesis. Only specific subpopulations of the breast cancer cells were tumorigenic, most likely corresponding to the cancer stem cells of the primary breast tumors tested. In this experiment, cancer stem cells were identified and isolated on the basis of the expression of specific surface markers. Consistent with the hypothesis that they indeed constitute the cancer stem cells of the tumor tested, they were able to grow tumors in mice even when about 100 cells were injected, contrary to the other cell subpopulations for which even tens of thousands of cells were incapable of forming tumors.

It should be mentioned that, despite the fact that they are very informative, such experiments are only indicative since formal proof of the fact that such cell populations constitute the cancer stem cells would only be derived by performing specific marking experiments and assessing whether the cells under in-

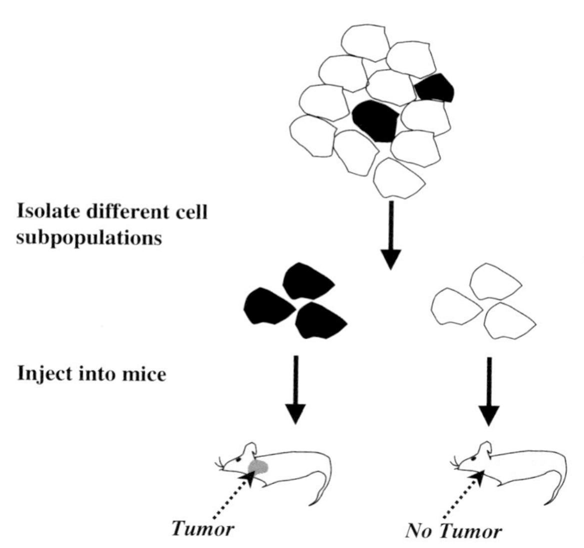

Fig. 2 Identification of the subpopulation of cancer stem cells requires the isolation of different subtypes of the cells within a tumor and assessment of their tumorigenicity in mice. Only cancer stem cells have the ability to grow in mice, whereas the "differentiated" cancer cells do not.

vestigation were sufficient to generate the full spectrum of cancer cells that constitute the tumor.

Another aspect of the stem cell theory is the contribution of the microenvironment to the particular fate that is going to be followed by those cancer stem cells. This became apparent from experiments performed almost 3 decades ago demonstrating that embryonal carcinoma cells injected subcutaneously form teratocarcinomas, while when injected into blastocysts they give rise to normal chimeric mice (Fig. 3). Thus, cancer stem cells, like their normal counterparts, strongly depend on their microenvironment for the acquisition of particular cell fates.

Various signaling pathways are expected to play important roles in cancer stem cell biology. Examples are offered by the Notch and *wnt* signaling pathways that both have been identified as potent regulators of stem cell fate under normal conditions. Interestingly, both represent potent oncogenes for breast and other epithelial tissues.

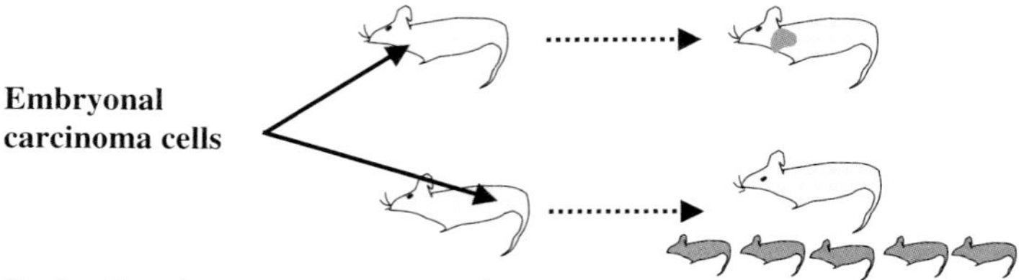

Fig. 3 That the microenvironment plays an important role in carcinogenesis is exemplified by the finding that injection of embryonal carcinoma cells subcutaneously results in the development of tumors, whereas injection in the blastocysts results in the development of normal chimeric mice.

Implications of Cancer Stem Cells in the Prognosis and Therapy of Cancer

Consistent with the multistage nature of carcinogenesis and the progressive accumulation of mutations by cancer cells, in "cancer stem cell" theory the cells that when mutated are the ones that affect the behavior of a tumor must be the cancer stem cells. Mutations in the "differentiated" cancer cells, no matter how dramatically they affect their properties, will be lost due to their limited lifespan. On the contrary, mutations in the stem cells will be inherited into their progeny and subsequently will contribute to the tumor's characteristics indefinitely, depending on the proliferative advantage they may offer.

Thus, there must be some heterogeneity even within the cancer stem cells of the same tumor. Indeed, when different types of breast cancers were assessed for the type and fraction of stem cells it was found that in a more aggressive form of the disease not only a higher fraction of cells was tumorigenic (thus representing the stem cells), but also this fraction could be subdivided into different subfractions, reflecting mutational events in the cancer stem cells. It is also conceivable that a mutation in a "differentiated" cell can also result in the acquisition of self-renewal properties, i.e. the generation of a cancer stem cell by a "differentiated" parental cell.

Therefore, the overall rate of tumor growth not only reflects the rate of cell proliferation, and the efficiency of the interaction between the neoplastic and the stromal component of a tumor, but also the proportion of cancer stem cells within a tumor (Fig. 4). This is supported by the observation that more aggressive tumors generally contain increased number of stem cells. For example, breast cancers of the "comedo" type (an aggressive form of the disease) can contain more than 60% of stem cells, as compared to less aggressive types of the disease that may contain cancer stem cells at a ratio as low as 1%, as shown for some pediatric astrocytomas.

The stem cell theory of cancer has some very important implications for the therapy of the disease. In principle, the major aim of targeted cancer therapy is to selectively inhibit the growth of cancer cells and to leave the rest of the organism unaffected. This selectivity is achieved by the identification of specific mole-

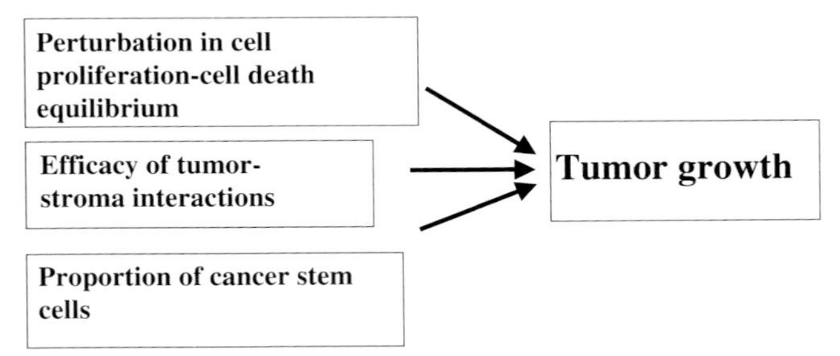

Fig. 4 In addition to the imbalance between cell proliferation over death, and the efficiency of the interaction between cancer and stromal cells (including vascularization), the overall rate of tumor growth is also determined by the content of cancer stem cells in the tumor.

cules in (or on) the cancer cells that will either constitute the exact targets of a given therapeutic approach or, alternatively, will serve as "recognition" molecules that will identify and distinguish the neoplastic cells for which the delivery of the specific therapeutic agents has been designed. However, this selectivity usually refers only to the vast majority of the cancer cells that constitute the tumor and correspond to the "differentiated" cells, ignoring the profile of the stem cells that are the ones that should constitute the target for an efficient therapeutic strategy. Given the differences in the properties of the cancer stem cells and their "differentiated" progeny, it is conceivable for these cell populations to differentially express these "recognition" molecules, thus rendering targeted therapy inefficient. The fact that stem cells represent only a small fraction of the tumor renders the identification of their expression profile more complex, as it is usually masked by that of the "differentiated" cells within the tumor. Furthermore, cancer stem cells are usually quiescent, possessing increased DNA repair capabilities and thus exhibiting higher drug resistance than the "differentiated" cells. Finally, cancer stem cells, like normal stem cells, express at high levels some transporter proteins, such as the ABC (ATP binding cassette), that have been recognized to play an important role in the molecular mechanism of anticancer drug resistance.

On the basis of this hypothesis it is conceivable that the partial (in)efficiency of conventional anticancer therapies is due to the fact that while they are successful in inhibiting the growth of a subset of the cancer cells, most likely that of the "differentiated" cells, they leave the stem cells virtually unaffected. Consistent with this notion, the occasional success of repeated treatments may be due to the fact that they manage to periodically diminish the population of "differentiated" cells. Thus, the development of a highly successful anticancer therapy with the potency to cure instead of managing the disease should also take into account that it is the stem cell population of a tumor that is the one that should be eliminated.

Bibliography

Al-Hajj M, Clarke MF. Self-renewal and solid tumor stem cells. *Oncogene* **2004**, *23*, 7274–7282.

Al-Hajj M, Wicha MS, Benito-Hernandez A, Morrison SJ, Clarke MF. Prospective identification of tumorigenic breast cancer cells. *Proc Natl Acad Sci USA* **2003**, *100*, 3983–3988.

Bissell MJ, LaBarge MA. Context, tissue plasticity, and cancer: are tumor stem cells also regulated by the microenvironment? *Cancer Cell* **2005**, *7*, 17–23.

Dean M, Fojo T, Bates S. Tumor stem cells and drug resistance. *Nat Rev Cancer* **2005**, *5*, 275–284.

Jones RJ, Matsui WH, Smith BD. Cancer stem cells: are we missing the target? *J Natl Cancer Inst* **2004**, *96*, 583–585.

13
Determination of Therapeutic Efficacy – Pharmacogenomics

From the very early days of anticancer therapy oncologists have observed that tumors, even in the same individual, are characterized by extremely high heterogeneity with regard to their response to treatment. Although some tumors respond very well to therapy, others, frequently even of the same type and classification, do not. Even more interesting and challenging at the same time is the fact that most tumors, while initially they may respond to the therapy to some degree, acquire resistance almost deterministically following repeated application. Thus, practically current therapy, by definition, is not destined to cure cancer, but rather to impede the progression of the disease.

When trying to understand the molecular determinants for the efficacy of anticancer therapy, which are reflected in the molecular mechanism(s) of the acquisition of resistance, one should always keep in mind that a very wide variety of drugs against cancer are available and each one of them targets specific pathways, thus employing different means to fight the disease. Therefore, different mechanisms can also account for rendering them more or less effective against cancer cells and, with certain exceptions, resistance to one treatment moiety does not necessarily mean reduced sensitivity against another. In other cases, however, where treatment targets more general and common pathways, the acquisition of resistance spans different strategies. For example, both chemotherapy with drugs, such as adriamycin and *cis*-platinum, and radiation aim to treat cancer by inducing, at least in part, apoptosis in the cancer cells. Therefore, the therapeutic efficacy of the treatment is proportional to the degree to which apoptotic cell death is induced. Thus, it is conceivable, and indeed experimentally proven, that defects in the cells' apoptotic machinery usually cause resistance to apoptosis-based therapy. This possibility is also applicable for cases such as the hormonal treatment of gynecological tumors. In these cases, a primary (and actually predictable) cause of resistance is the absence of the corresponding hormonal receptors. However, even if they are present, the hormone ablation of the tumor aims to "starve" the tumor from a major mitogen, finally causing apoptotic cell death. Thus, again an efficient apoptotic response is a prerequisite for effective treatment. Another mechanism that may also account for the reduced sensitivity of cancer cells to treatment is the presence of specific drug pumps, such as P-glycoprotein that clears the drug from the cells. Impor-

Understanding Carcinogenesis. Hippokratis Kiaris

ISBN 3-527-31486-5

tantly, a common observation is that cancer cells frequently overexpress these cellular detoxifiers.

We will discuss below some of the aspects of the mechanisms that may confer resistance/sensitivity to anticancer therapy with special emphasis given to apoptosis – the predominant mechanism by which anticancer drugs produce their effects. The importance of this mechanism for resistance acquisition as compared to other mechanisms is also due to the fact that defects in apoptotic pathways are intrinsic to cancer cells – linking an endogenous property of the malignancy with the cause of reduced therapeutic efficacy. Finally, we will also introduce the field of pharmacogenomics and the emerging concept of individualized therapy. Contrary to the molecular determinants of anticancer therapeutic efficacy in cancer cells, pharmacogenomics refers to inherited polymorphisms that are not directly associated with the tumors' cellular physiology, thus producing a phenotype only in association with the use of specific drugs.

Intrinsic Link between Chemotherapy/Radiotherapy and Apoptosis

Apoptosis or programmed cell death is defined by certain morphological changes in which caspases play an important role. It is an active process that is precisely controlled and results in the death of the cell in a controlled manner. It is triggered by various signals that are either exogenous (such as genotoxic stress) or endogenous during certain developmental and physiological stages. Most anticancer drugs have been identified during empirical screens on the basis of their potency to inhibit (cancer) cell proliferation and thus their precise mechanism of action remains obscure in various cases. Frequently, however, it became apparent that they act, at least in part, by a mechanism that involves the initiation of endogenous apoptotic programs.

Apoptosis, to the extent that caspase activation is involved, is triggered by extrinsic and intrinsic pathways. The extrinsic pathway involves the death receptors [CD95, tumor necrosis factor (TNF) receptor and TNF-related apoptosis-inducing ligand (TRAIL) receptor] that upon activation and via a series of sequential events of caspase activation result in the activation of certain caspases, such as caspase-3 and -7. For the intrinsic pathway, mitochondrial proteins such as Smac/DIABLO (second mitochondria-derived activator of caspase/direct inhibitor of apoptosis-binding protein with low p*I*), HtrA2 (high-temperature requirement protein A2) and cytochrome *c* are released. Cytochrome *c* in association with apoptotic protease-activating factor (Apaf)-1 activate caspase-9, inducing apoptosis, while Smac/DIABLO and HtrA2 bind and inhibit proapoptotic proteins, suppressing cell death.

p53 Tumor Suppressor: A Major Regulator of Anticancer Therapeutic Efficacy

A key regulatory role for the apoptotic process is reserved for the tumor-suppressor gene (TSG) p53. p53 can alter the equilibrium between proapoptotic (Bax, Bac, Puma, etc.) and antiapoptotic (Bcl-2) genes in favor of the first, and can induce other genes implicated in apoptosis such as PTEN (phosphatase and tensin homolog deleted on chromosome 10) and Apaf-1, while it may also stimulate the extrinsic pathway for caspase activation mentioned before by activating CD95 and TRAIL receptor 2.

That cancer cells must evade apoptosis is common ground in tumorigenesis. Thus, considering the importance of p53 in regulating apoptosis, it is not surprising that p53 mutations represent the most common genetic lesion in human cancers.

The importance of various genes regulating apoptosis, including p53, in determining the efficacy of various anticancer drugs has been shown by several studies in mice and *in vitro* in which the sensitivity of tumors (or cells) to certain apoptosis-inducing anticancer agents has been restored following re-expression of the defective genes, or the reverse experiments that showed that inactivation of these genes was sufficient to render tumors resistant to chemotherapy and/or radiotherapy. Thus, in general, an intact apoptotic response is needed for efficient anticancer therapy.

With regard to the application of ionizing radiation, a common and usually quite successful anticancer approach, apoptosis in general is not considered as the predominant response of the tumor against the therapy. In that case, following the exposure to ionizing radiation the majority of the cancer cells undergo mitotic catastrophe, the condition that characterizes the result of a series of successive aberrant mitoses, and eventually cell death or permanent (irreversible) growth arrest.

However, whilst the acquisition of resistance to drug-induced apoptosis is currently considered as the consensus with regard to how resistance to therapy is regulated, it has to be noted that certain experimental evidence challenges this notion and implies a mechanism that at least superficially may be viewed as the opposite, thus complicating the picture. For example, it has been demonstrated that the presence of $p21^{Waf1}$, a TSG gene that acts downstream of p53 as its effector in driving the cell cycle arrest, renders tumors resistant to anticancer therapy. Thus, according to these findings, an intact p53 response results in resistance instead of sensitivity to therapy. Most likely this is associated with the ability of $p21^{Waf1}$-expressing cells to stop dividing in the presence of the stress induced by the therapy and thus "repair" or circumvent the damage caused by it, while cells deficient to $p21^{Waf1}$ continue to proliferate, accumulate the damage caused by the therapy and finally die. Of course, in this case another aspect of anticancer therapy is implied – that involving the induction of cellular senescence, in contrast to apoptosis as mentioned earlier. In this case, therapy may induce senescence, which in turn is capable of stimulating tumor growth by nonautonomous mechanism(s). Note that the same group of investigators, led by B. Vogelstein, also reported that the

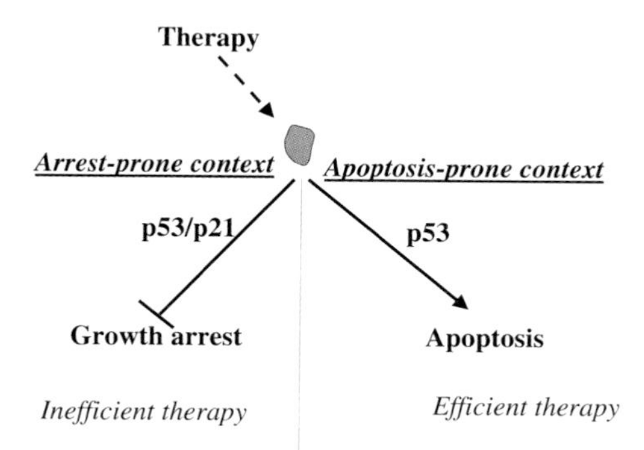

Fig. 1 In terms of the efficacy of anticancer therapy, the dual role of the p53/p21 pathway is probably related to the intrinsic susceptibility of certain cell types to undergo cell cycle arrest instead of apoptosis and *vice versa*.

same p53 tumor suppressor offers resistance against certain DNA intercalating agent-based drugs, while it confers sensitivity against others such as 5-fluorouracil (5-FU) that act by a different mechanism (see below).

This apparent controversy most likely reflects the multiple roles of p53 in regulating apoptosis, genomic stability and growth arrest (reversible and irreversible). The origin of cancer cells in combination with specific defects in certain pathways related to p53 may favor one route versus another, thus resulting in the contradictory role of p53 in the prediction of response to anticancer therapy. In other words, it is likely that in cell types which are more prone to enter (irreversible?) growth arrest following exposure to genotoxic stress (anticancer treatment), the intact p53/p21-related response confers resistance, whereas p53 confers sensitivity to the therapy in cells that are more prone to apoptotic cell death (Fig. 1). Indeed, as discussed earlier in the corresponding chapters (Chapters 4 and 11), the same signal that can induce, notably by the same molecular circuit, apoptotic cell death in a given cellular context, can cause cell cycle arrest in a different context. This underlines the importance of the cellular context, in addition to the genetic status as important determinants for anticancer drug efficacy.

Resistance to Therapy is Intrinsic to the Tumor

From the aforementioned findings it becomes apparent that the ability of cancer cells to evade apoptosis is also responsible for rendering tumors resistant to therapy. Thus, resistance is intrinsic to the tumors and not a property acquired during carcinogenesis after cancer cells are subjected to a form of anticancer therapy. Of course, genetic heterogeneity increases during tumor progression and selection during treatment may favor certain genotypes that display increased resistance to the therapy as compared to others. Consistent with this notion is the observation that tumor metastases, that by definition consist of cells already subjected to a first intense selection "bottleneck" for surviving under unfavorable conditions (see Chapter 10), are generally more resistant to therapy than primary tumors. This is also supported by the enrichment of metastatic tumors for p53 mutations and Bcl-2 overexpression as compared to the primary tumors from which they originated. The relatively higher success of combination therapy consisting of more than one agent with distinct targets is based on exactly this reduced probability of cancer cells having intrinsic resistance to these targets.

Pharmacogenetics

Until now we have focused on molecular pathways in the cancer cells that determine how efficient cancer therapy is going to be. In this case the variability in the response is largely due to *de novo* mutations that have accumulated in the cancer cells, which constitute exactly the same targets of the anticancer therapy. However, inherited polymorphisms in the patient may also play a considerable role in determining the efficacy of anticancer therapy.

The term pharmacogenomics used to describe a distinct scientific discipline was introduced in 1959 and referred to the exploration of the inherited basis of drug-response phenotypes. While initially it aimed to explain the variability in the response to drugs against infectious agents, such as malaria, it soon became apparent that it could be extended to anticancer drugs as well. It has to be emphasized that such polymorphisms, in the coding or regulatory regions of certain genes, affecting the outcome of certain therapies, are quite common in the human genome considering that such single nucleotide polymorphisms exist every 1000–3000 bp. The mechanisms that may determine the efficacy by which a drug will act are quite divergent. They can be categorized into three groups based on whether they act upstream of, at or downstream of the critical targets (Fig. 2).

Several such examples are available in the scientific literature and some of them will be discussed below.

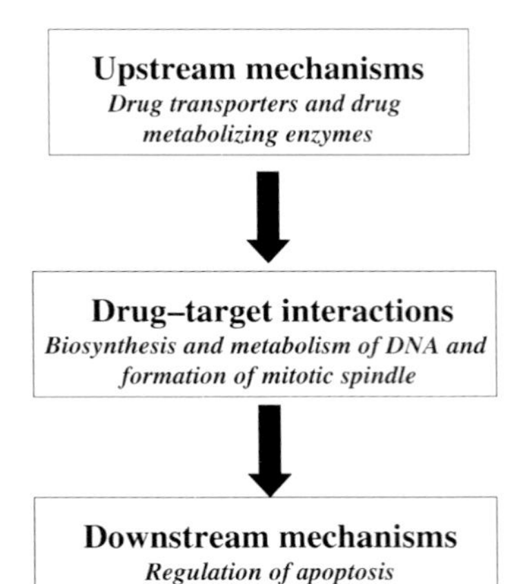

Fig. 2 The three different levels at which resistance to therapy may be encountered (see text for more details).

Upstream Mechanisms

In general, this category includes drug transporters responsible for uptake or excretion, carriers, and drug-metabolizing enzymes that are implicated in the activation, inactivation and detoxification of the drugs.

ABC (ATP-binding cassette) transporters, including the multidrug-resistance gene MDR1 encoding for P-glycoprotein, are also thought to play an important role in determining anticancer drug efficacy by regulating the amount of drug to be inserted in the cancer cell. This family of genes, that in addition to MDR1 also includes MRP1 and MRP2, are usually targeted by amplification in the cancer cells (somatic mutations), but are also characterized by inherited polymorphisms that affect the outcome of therapy.

Cytochrome P450 monoxygenases are enzymes that can either detoxify drugs such as paclitaxel and tamoxifen or activate inactive prodrugs such as cyclophosphamide. Paclitaxel is metabolized by CYP2C8 that has different alleles with distinct efficiency to metabolize this drug. CYP17 is a member of the cytochrome P450 enzymes, consisting of enzymes responsible for the oxidative metabolism of endogenous hormones and steroids. CYP17, in particular, is involved in the biosynthesis of estrogens. A specific polymorphism in the untranslated region of the CYP17 gene results in its upregulation by an Sp1-dependent mechanism that, in turn, affects the pace at which estrogens are synthesized. It has been found that the risk for endometrial cancer development in estrogen replacement therapy greatly depends on the specific CYP17 allele carried by the patient

UDP-glucuronsyltransferases (UGTs) catalyze the glucuronidation of many xenobiotics and endobiotics in order to increase their water solubility and thus facilitate their elimination. The prodrug irinotecan is converted in the liver into a DNA topoisomerase I inhibitor that is used against colon cancer. Polymorphisms in the promoters of specific UGTs result in decreased transcription and thus reduced glucuronidation for the corresponding targets, including the active drug mentioned above. The latter in some cases may induce diarrhea-related toxicity.

Drug–Target Interaction

5-FU is a widely prescribed drug that is used against various solid tumors. Upon introduction into the organism it is converted into 5-fluoro-2-deoxyuridine monophosphate, an inhibitor of thymidylate synthase (TS), which suppresses cell division. In the liver a large portion of 5-FU is inactivated by dihydropyrimidine dehydrogenase (DPD) – an enzyme that exhibits high variability in activity among individuals. This variability ranges between 8- and 21-fold, and is attributed to common polymorphisms in its genomic sequence. In general, individuals with reduced DPD activity display high toxicity against 5-FU therapy. Polymorphisms in the regulatory region of TS that affect the levels of TS expression and thus its overall activity also play an important role in the determination of the efficacy of 5-FU therapy. Lower TS activity has been correlated with a better response against 5-FU. Thus, determination of the polymorphisms in TS and DPD may predict the outcome against 5-FU therapy.

DNA repair enzymes that correct damage induced by some anticancer agents are also thought to play a role in the efficiency of these drugs. For example, a polymorphism at −6 exon 13 T → C of the DNA mismatch repair gene hMSH2 was found to modulate the susceptibility to O^6-guanine-alkylating chemotherapy.

Antifolates such as methotrexate are used against leukemias, lymphomas and breast cancer in order to interfere with the metabolism of folate that is involved in amino acid and nucleotide biosynthesis. The enzyme 5,10-methylenetetrahydrofolate reductase, which plays a crucial role in the maintenance of reduced folate levels, exhibits polymorphisms that affect its activity and thus the levels of endogenous folate.

Downstream Mechanisms

Downstream mechanisms include the pathways that regulate the induction of apoptosis. Their role in anticancer therapy has been discussed in detail previously, especially in view of the fact that these genes are frequently altered during carcinogenesis. However, somatic mutations in cancer cells are not the only genetic mechanism that modulates their activity. Certain polymorphisms between individuals have also been described in the scientific literature that potentially contribute to the determination of the outcome of anticancer therapy. For example, polymorphisms in the p53 tumor suppressor have been reported that affect its activity – an effect that, in turn, modulates susceptibility to therapy. In exon 4, a polymorphism encoding either for Arg or Pro at position 72 has been reported that affects the efficacy of *cis*-platinum-based therapy in head and neck cancer patients.

Finally, we have to mention that polymorphisms in the mitochondrial DNA may also modulate susceptibility to anticancer therapy. A recent example involves the identification of a linkage between the mitochondrial haplogroup *J* and *cis*-platinum-induced hearing impairment.

Towards Individualization of Cancer Therapy

Pharmacogenetics aims to predict the outcome of therapy by considering the patient's individual genotype. The recent publication of the human genome revealed that great variation exists between individuals, and this remains to be recorded and evaluated. It is anticipated that a subset of this variation will affect the efficacy of anticancer therapy. Ongoing, large-scale studies are scoring the incidence of polymorphisms not only within the coding, but also in the noncoding, potentially regulatory sequence of genes that encode for potential regulators of anticancer drug efficacy. It is expected that in the near future progress in this field will permit the selection of specific drugs to be used in association with the patient's genotype.

Bibliography

Bunz F, Hwang PM, Torrance C, Waldman T, Zhang Y, Dillehay L, Williams J, Lengauer C, Kinzler KW, Vogelstein B. Disruption of p53 in human cancer cells alters the responses to therapeutic agents. *J Clin Invest* **1999**, *104*, 263–269.

Efferth T, Volm M. Pharmacogenetics for individualized cancer chemotherapy. *Pharmacol Ther* **2005**, *107*, 155–176.

Johnstone RW, Ruefli AA, Lowe SW. Apoptosis: a link between cancer genetics and chemotherapy. *Cell* **2002**, *108*, 153–164.

Komarova EA, Gudkoc AV. The role of p53 in determining sensitivity to radiotherapy. *Nat Rev Cancer* **2003**, *3*, 117–129.

Relling MV, Dervieux T. Pharmacogenetics and cancer therapy. *Nat Rev Cancer* **2001**, *1*, 99–108.

Schmitt CA. Senescence, apoptosis and therapy: cutting the lifelines of cancer. *Nat Rev Cancer* **2003**, *3*, 286–296.

Waldman T, Zhang Y, Dillehay L, Yu J, Kinzler K, Vogelstein B, Williams J. Cell-cycle arrest versus cell death in cancer therapy. *Nat Med* **1997**, *3*, 1034–1036.

14
Certain Chemicals Induce Cancer: Chemical Carcinogenesis

That certain chemicals can induce cancer (more importantly, specific forms of the disease) has been well established for many years. Classical examples include the development of scrotal cancer in chimney sweeps nearly 250 years ago. In this case, the incidence of scrotal cancer decreased substantially following the recommendation of the authorities for frequent bathing. As in any other case where carcinogenesis was induced by exposure to certain chemicals, it became apparent that continuous (chronic) exposure to the harmful substances associated with the occupation of the chimney sweeps was essential for the onset of neoplasia that required rather increased latency and developed late in life.

Other well-known examples of carcinogenesis associated with exposure to certain carcinogens include the development of mesothelioma by asbestos workers and the development of lung cancer in tobacco smokers.

In chemically induced carcinogenesis the compound that is responsible for the induction of neoplasia damages the DNA, resulting in the development of mutations that exhibit a striking specificity depending on the specific chemical. These mutations (e.g. DNA breaks, base substitution, insertions or deletions, crosslinking of the two DNA strands, etc.) should bypass the mechanisms that correct such DNA

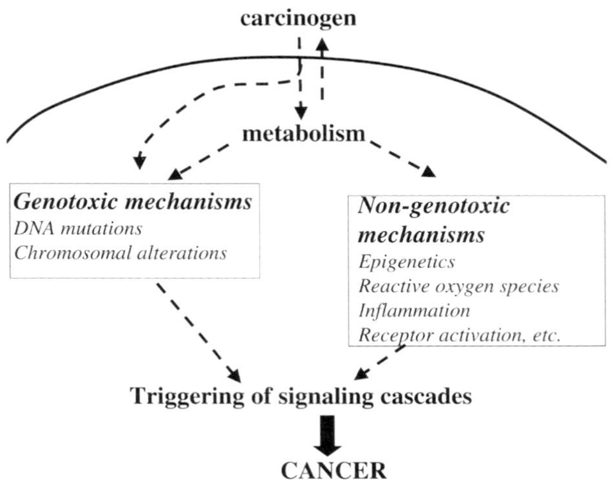

Fig. 1 Genotoxic and nongenotoxic chemical carcinogenesis. In the first case, the carcinogen induces certain mutations in the DNA that in turn alter the expression profile of the cells; in the second case, DNA remains unaltered by the carcinogen.

Understanding Carcinogenesis. Hippokratis Kiaris

ISBN 3-527-31486-5

damage, escape DNA repair mechanisms and be fixed in the genome. Usual targets of such mutations in the DNA ultimately include oncogenes and tumor-suppressor genes (TSGs) that are activated or inactivated, respectively. Thus, the mutations induced by the chemical carcinogens provide the genetic hits that predispose the cell to malignant transformation. However, there are also many carcinogens that are not able to bind to the DNA and, thus, they exert their oncogenic properties by mechanisms that do not directly involve mutagenicity (Fig. 1).

We will now describe the mode of action of some carcinogens that may be considered representative for chemical carcinogenesis.

Direct DNA Binding or Metabolic Activation of Some Carcinogens – Genotoxic Carcinogens

Direct binding of some chemicals to DNA and the subsequent induction of certain alterations in its primary structure represents an apparent, "straightforward" mechanism for carcinogenesis induced by specific chemicals. This class of carcinogens, also called "direct carcinogens", includes ethylene oxide, bis(chloromethyl)ether and some aziridine or nitrogen mustard derivatives. These compounds are electrophilic, having the ability to physically interact with the DNA to induce mutations.

However, the vast majority of the chemicals that are known to act as chemical carcinogens do not have the ability to interact directly with DNA, but need certain metabolic modifications (activation). Thus, they are considered as procarcinogens. The major role in the conversion of certain procarcinogens into activated carcinogens is played by the cytochrome P450 multienzyme complex and the P450-related monoxygenases. To date, more than 50 genes encoding P450-associated enzymes have been identified that exhibit tissue-specific expression and overlapping substrate specificity.

Polycyclic Aromatic Hydrocarbons (PAHs)

PAHs are quite common carcinogens that are produced during the combustion of virtually any organic material. An example of a PAH is the pentacyclic benzo[*a*]pyrene (BP) that is present in the coal tar. BP, like other PAHs, as well as other carcinogens, either becomes excreted or metabolically activated upon entry into the cell. BP is transformed into the 7,8-dihydrodiol of BP that has been shown able to directly bind to DNA. This process involves many enzymes, with CYP enzymes playing an important role. The importance of DNA binding in the carcinogenic properties of the metabolically activated PAHs and other carcinogens becomes apparent from the fact that there is a correlation between the carcinogenic properties of such chemicals and the affinity by which they bind to DNA. To date, several dyes that can bind to DNA are considered as potent carcinogens. Important intermediate metabolites of BP as well as other PAHs are epoxides, with diol-epoxides being recognized as the ultimate metabolites that mediate PAH-induced tumorigenesis.

Aflatoxins

Aflatoxins belong to the mycotoxins that contaminate food during the growth of certain molds. Aflatoxins, particularly aflatoxin B1 (AFB1), were found to be responsible for the development of hepatocellular carcinoma. AFB1 requires P450 activation and its carcinogenic (mutagenic) properties are also exerted through the epoxidation pathway. An important aspect of this activation is the stereochemistry of the mutagenic metabolites. For example, CYP3A4 that mediates the activation of the AFB1 in the liver exclusively produces the exo-isomer. In this case, the endo-isomer is less than 1000-fold genotoxic.

Nongenotoxic Carcinogens

Some chemicals do not have the ability to directly bind to DNA and, thus, are not mutagenic, yet they are potent carcinogens in animals.

2,3,7,8-Tetrachlorodibenzo-*p*-dioxin (TCDD) belongs to this category of carcinogens. TCDD is generated during the manufacture of polychlorinated phenols. Epidemiological studies in humans and experimental studies in animals have associated TCDD with the development of various cancers, such as lung, liver, skin and other cancers. In contrast to the PAHs that require metabolic activation, the mechanism by which TCDD is thought to produce its carcinogenic effects involves the direct agonistic stimulation of the arylhydrocarbon receptor (AhR) that through a cascade of signaling events results in the modulation of activity of proteins that regulate cell differentiation, proliferation and death. Consistent with the oncogenic effects of TCDD, genetic ablation of the Ah gene in mice renders animals resistant to TCDD-induced carcinogenesis while overexpression of an activated form of AhR induces cancer in mice. Indeed, microarray analysis shows that exposure to TCDD results in the differential expression of a wide variety of endogenous genes (more than a few hundred), many of which have been shown previously to be linked with carcinogenesis in a causative manner.

Bibliography

Klaunig JE, Kamendulis LM. The role of oxidative stress in carcinogenesis. *Annu Rev Pharmacol Toxicol* **2004**, *44*, 239–267.

Luch A. Nature and nurture – lessons from chemical carcinogenesis. *Nat Rev Cancer* **2005**, *5*, 113–125.

Poirier MC. Chemical induced DNA damage and human cancer risk. *Nat Rev Cancer* **2004**, *4*, 630–637.

Wogan GN, Hecht SS, Felton JS, Conney AH, Loeb LA. Environmental and chemical carcinogenesis. *Semin Cancer Biol* **2004**, *14*, 473–486.

15
Hormones and Cancer

The link between cancer and hormones has been recognized for more than a century, and it also constitutes a defined area in research and clinical practice termed endocrine oncology. This intimate linkage between endocrinology and oncology reflects the fact that hormones are directly or indirectly involved with various aspects of tumor biology ranging from the basic mechanisms of carcinogenesis to everyday oncological practice. For example, hormones are important regulators of cancer growth, producing either negative or positive effects on cell proliferation. Furthermore, the hormonal status of a tumor possesses important prognostic significance and implies the potential applicability of therapeutic strategies to be followed. As an example we mention the acquisition of androgen independency by prostatic tumors that accompanies their metastatic behavior. Finally, hormones or artificial hormone analogs are used as anticancer drugs and continue to be intensely developed, e.g. the various antiestrogens used against breast cancer.

Before we describe in some detail the action of selected hormones in cancer growth, we have to keep in mind that at the level of tumor biology hormones generally produce their effects by "systemic", "paracrine" or "autocrine" mechanisms (Fig. 1). This classification refers to the action of hormones from the endocrine point of view and describes the following three possibilities. (i) "Systemic" action is where the hormone is produced by a central organ or tissue and through systemic circulation reaches the target organs where it mediates its effect(s) through specific receptors. (ii) "Paracrine" action refers to the production of a hormone by

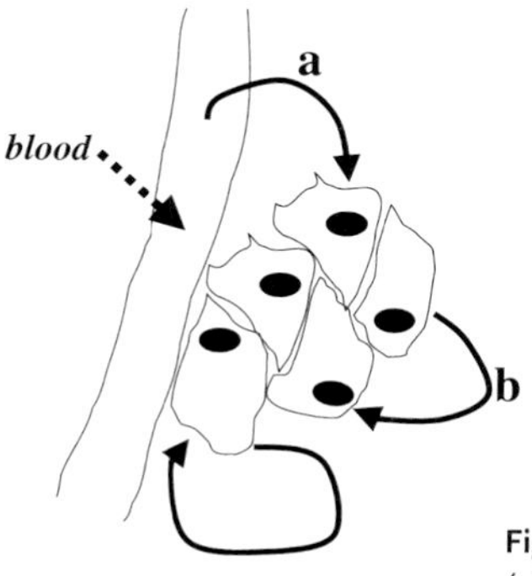

Fig. 1 (a) Systemic, (b) paracrine and (c) autocrine mechanisms of hormone action.

Understanding Carcinogenesis. Hippokratis Kiaris

ISBN 3-527-31486-5

a cell that (the hormone) subsequently acts onto a cell physically located close to the cell that produced it. This type of hormone action usually mediates the communication of different cell types located close to each other. (iii) "Autocrine" action covers the cases where both the producing cell and the target cell are the same. This latter action is of particular importance in carcinogenesis, for which we evidence the formation or the intensification of (pre-existing) autocrine loops. In drug development, the disruption of such loops usually constitutes the desired target in tumor growth inhibition.

Considering that hormones essentially constitute signaling molecules, at the molecular level, they act by mechanisms already seen in various signaling pathways. Thus, they usually act through specific receptors that can be located in the cell membrane and are linked to other signaling molecules that trigger a signaling cascade upon hormone binding and receptor activation. Hormones that typically act through this mechanism are peptide and protein hormones. Alternatively, steroid hormones act (predominantly as we will see below) in the nucleus, activating their receptors, which can function as transcription factors (Fig. 2).

The production of opposing results is something that we frequently see in the action of the same hormone in different tissues and is the consequence of the involvement of certain tissue-specific cofactors that modulate the hormone's activity. Therefore, formally speaking, it is difficult to attribute a certain causative associa-

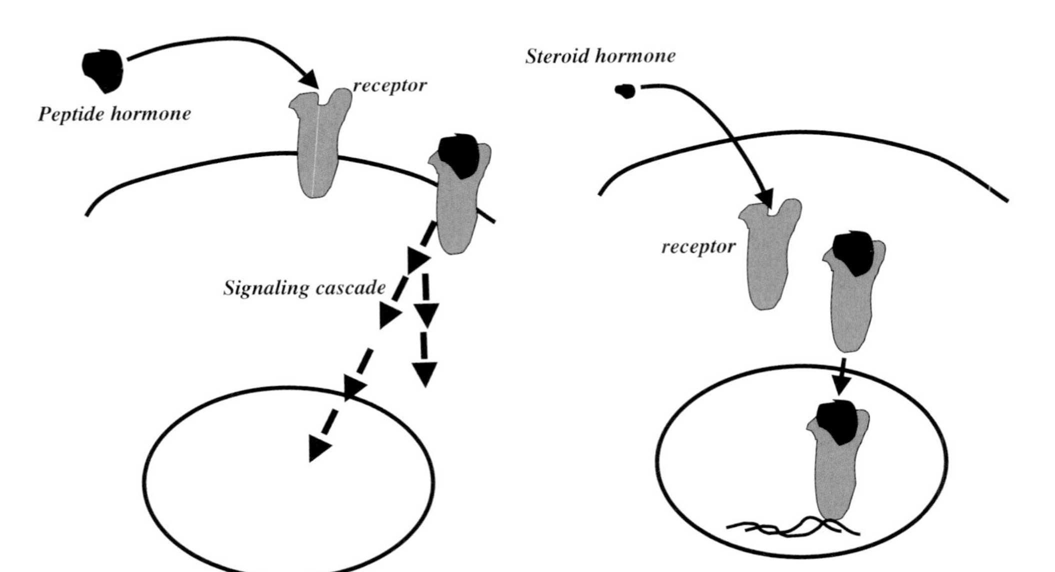

Fig. 2 General mechanism of action of peptide/steroid hormones. In principle, peptide hormones (left) act through membrane receptors that upon activation trigger signaling cascades that finally may reach the nucleus, thus altering the expression profile of the recipient cells. Steroid hormones (right) act predominantly through receptors that upon binding translocate into the nucleus, thus operating as transcription factors.

tion between specific hormones and biological effects unless we refer to a given cellular context.

We will now describe in some detail the action and role in carcinogenesis of selected hormones that cover a wide range of the possible mechanisms by which they may regulate neoplastic growth. These hormones are the sex steroids that are implicated in breast and prostate cancer, and exert their action predominantly in the nucleus, and growth hormone-releasing hormone (GHRH) that is a short peptide hormone implicated in various cancers via systemic as well as autocrine/paracrine mechanisms.

Androgens and Prostate Cancer

Prostate cancer is among the leading causes of cancer-related death in the Western world. That androgens play an important role in prostate cancer development has been suggested by three lines of evidence: (i) castration of young males offers resistance to the development of prostate cancer, (ii) epidemiological studies imply that there is a clear correlation between the levels of androgens in the serum and the risk of developing of prostatic tumors, and (iii) *in vitro* and *in vivo* experimental systems show that the equilibrium of cell death versus proliferation can be altered by the removal/supplementation of androgens.

The major androgen in the serum is testosterone that is produced in the Leydig cells in the testis. In the prostate, the most important androgen is 5α-dihydrotestosterone (DHT) that is converted from testosterone by 5α-reductase. Both androgens act through the same receptor – the androgen receptor (AR) – that is the key player in prostate cancer. AR is a nuclear receptor that upon binding to the androgens functions as a transcription factor in a ligand-dependent manner. The activation of AR by the corresponding ligands results in the dissociation of the chaperone proteins that are bound to the inactive receptor, its dimerization and, finally, its binding to the corresponding responsive element(s) in the DNA. This binding drives the transcriptional activation of target genes that initiate a signaling cascade that results in the stimulation of cell proliferation and inhibition of apoptotic cell death in the prostate, thus favoring neoplastic growth. In addition, stimulation of lipogenesis, through the activation of specific lipogenic transcription factors, is recognized as a result of AR activation, and serves in the synthesis of key membrane components such as phospholipids and cholesterol. Importantly, these lipogenic pathways persist from the earliest stages of prostate cancer, exemplified by prostatic intraepithelial neoplasia (PIN), through to the onset of androgen-independent disease, suggesting the fundamental importance of this process in prostatic carcinogenesis. In the absence of ligand, activation of the AR has also been reported and takes place through crosstalk with divergent growth factor pathways, such as those of the epidermal growth factor (EGF), insulin-like growth factor (IGF)-I, keratinocyte growth factor (KGF/FGF-7), etc.

Regarding the action of AR, a nongenomic mechanism has also been proposed and is thought to involve the activation of the mitogen-activated protein kinase

(MAPK) pathway through a c-Src-dependent mechanism. In a relevant model proposed recently the central role in this type of nongenomic AR signaling is played by the cholesterol-rich lipid rafts of the cancer cell membranes. The existence of an AR that is bound to the membrane has also been proposed, but remains to be identified.

Experiments in different systems, both *in vitro* and *in vivo*, revealed a high heterogeneity in the outcome (response) that follows AR activation. This heterogeneity is due to the fact that cell-specific factors can modulate the response to the receptor's activation. These "modifiers" of AR function have been termed coactivators or corepressors and are proteins that produce their biological effects through two major mechanisms: class I coactivators, such as p300/CBP that bridges the transcriptional machinery to the nuclear receptors, and class II coactivators, such as p160 that is involved in chromatin remodeling. According to a very simplified general model for the action of the coactivators, after receptor activation, subsequent dimerization and binding to the regulatory regions of the target genes, and depending on their relative availability in each particular cell type, coactivators are recruited alone (independently) or in combination to the transcription complex, followed by the recruitment of RNA polymerase II (Fig. 3). The transcriptional machinery that has now been built can bind to the DNA after appropriate chromatin remodeling that occurs after the action of specific coactivators. This mechanism is applicable not only for AR, but for the estrogen receptor (ER) as well. Thus, the specific effect of a given steroid hormone is not the simple result of the activation of the corresponding receptor, but is determined by the availability of a set of coactivators that are cell and tissue specific.

Various alterations in several coactivators of the AR have been studied in prostate cancer. For example, SRC-1 expression was found to be increased in specimens from patients that developed androgen-independent prostatic carcinoma, as compared to patients with androgen-dependent and benign prostatic lesions, whereas RAC-3 expression correlates with high tumor grade and stage. AR coactivators have also been found to be responsible for the androgen-independent activation of AR. For example, p300 has been implicated in the activation of AR, independently of androgens, by interleukin-6.

Various alterations in the AR have been reported in prostatic tumors, including mutations conferring increased activity and modulation of the ligand specificity. An important source for AR polymorphism targets the poly-Glu (Q) repeats, encoded by CAG triplets, in the N-terminal region of the AR gene. Polymorphism in the repetition of these triplets ranges from 14 to 35. The degree of repetition of

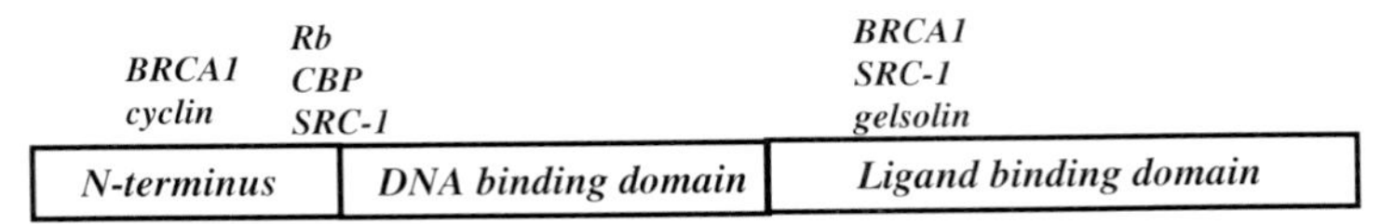

Fig. 3 Structure of the AR showing the major protein domains. The relative position of the coactivator-binding region is shown.

these CAG triplets has been associated with prostate cancer risk, in a manner whereby short repetitions increase the risk of prostatic carcinoma development and have also been linked with a younger age of diagnosis. Mutations in the AR are generally considered rare in prostatic carcinomas, at least prior to therapy. Amplification of AR that is located at Xq11–q13 is also uncommon in prostate cancer prior to therapy, but the incidence increases up to 30% (according to some investigators) in hormone-refractory tumors and in their metastases. According to this finding it is reasonable to speculate that AR amplification, which notably increases the levels of functional AR, offers a proliferative advantage in a hormone-depleted environment.

An important aspect of prostate cancer biology is the acquisition of androgen independency. After a certain point during prostate cancer development, prostatic carcinoma cells lose their sensitivity to androgens. Therefore, androgen ablation, a usual therapeutic approach for the management of prostate cancer, does not impede tumor growth any more. This point usually signifies the transition of prostate cancer into an incurable state. Importantly, at the molecular level this transition is not always associated with a loss or general alteration in the structure/expression of AR, as might have been expected, suggesting that other modifications are responsible for this phenomenon. Contrary to the functional deletion, amplification of AR has also been described above and may be associated with the "hypersensitization" of the cells to androgens (Fig. 4).

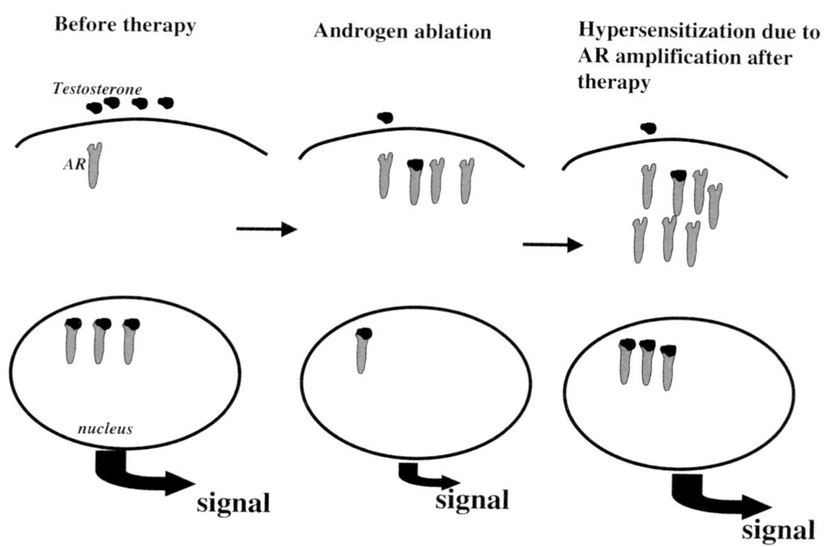

Fig. 4 Amplification of AR is frequently associated with hypersensitization of the cells to androgen. According to this model, overexpression of the intact AR protein compensates for the reduced levels of circulating androgens.

Sex Steroids and Breast Cancer

Breast cancer is another malignancy with a clear association with specific steroid hormones, especially 17β-estradiol. The role of estrogens, and especially of 17β-estradiol, in breast cancer is complicated by the action of another steroid hormone, i.e. progesterone, present in the mammary gland at considerable levels during specific phases of the menstrual cycle and may either suppress or induce mammary epithelial cell growth depending on the experimental system used.

The central role, in analogy with AR for prostate cancer, is played by ERs. Contrary to the existence of only one AR, two different ERs have been identified – ERα and ERβ. These receptors are encoded by genes located on chromosomes 6q and 14q, respectively. The corresponding receptors operate in a similar manner to that of AR. Thus, they contain domains with analogous functions, such as the DNA-binding domain, the transactivation domain, the ligand-binding domain, etc. Dimerization, that can be either homo- or heterodimerization between ERα and ERβ, is also essential for appropriate receptor function and transcriptional activation of the target genes. Regarding the molecular regulation of the ER-mediated action of sex steroids, in addition to the role of certain coactivators and corepressors that operate in a similar manner to that described for AR, an additional degree of complexity is contributed by the fact that the two types of ERs exhibit differential ligand-induced activity. Thus, their combined action may produce diverge results even in the same cell system.

Proper ER action is necessary in the mammary epithelium for normal gland development and maturation. In the normal breast only about 10% of the epithelial cells express ERα, contrary to about 85% of them that express ERβ. At the cell cycle level, it has been demonstrated that 17β-estradiol induces G_1/S transition resulting in the increase of the fraction of cells that are in the S phase. This is achieved by the ability of 17β-estradiol to regulate the expression of genes that can affect cell cycle progression, such as cyclin D1, *myc*, etc. (Fig. 5).

Modulation of ER expression is thought to play a significant role during the development of breast cancer; however, the precise mechanism remains obscure since contradictory data have been reported. In general, in invasive and *in situ*

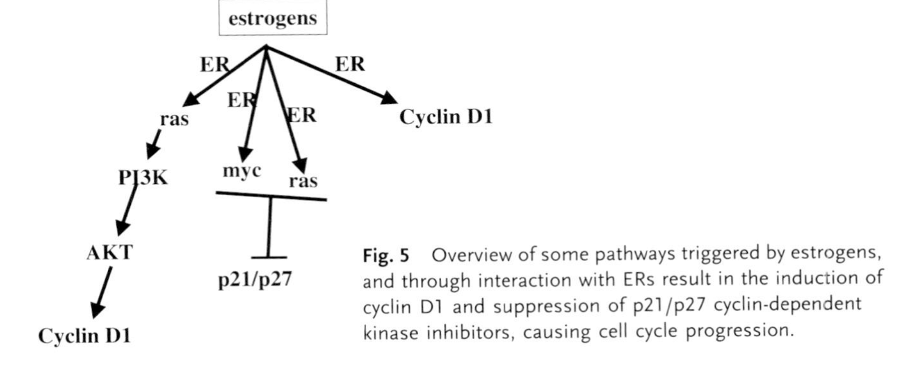

Fig. 5 Overview of some pathways triggered by estrogens, and through interaction with ERs result in the induction of cyclin D1 and suppression of p21/p27 cyclin-dependent kinase inhibitors, causing cell cycle progression.

breast cancer, the fraction of cells expressing ERα is thought to increase as compared to that obtained from normal mammary epithelium. ERβ expression decreases in malignant breast lesions and, thus, tumor-suppressor-like activity has been attributed to this particular type of ER. However, current opinion considers that it is not the net level of each individual receptor, but rather the ratio between ERα and ERβ that plays the deterministic role in breast cancer development, in a manner that predicts that estrogen dependency is caused by the increased ratio between ERα/ERβ, whereas a decrease in this ratio is consistent with the onset of estrogen-independent breast cancers. The picture is complicated further by the expression of specific splice variants of the ERs that possess altered signaling properties compared to those of the intact, full-length receptors. In addition to the altered expression, ERs in breast cancer specimens frequently undergo structural alterations that include mutations, deletions and amplifications.

However, regardless of the precise diagnostic and predictive significance of ERs in breast cancer, the importance of estrogen stimulation for the development of the disease has been well established since the late 19th century when G. Beatson showed that removal of the ovaries (i.e. surgical estrogen ablation) could be used successfully, notwithstanding with variable results, for the management of advanced breast cancer. Since that time, several attempts have been made to inhibit (antagonize) estrogen action by chemical agents. The results of those efforts were not very promising until the early 1970s when *tamoxifen* was developed. Tamoxifen is a nonsteroidal antiestrogen that can be used in patients for both the prevention and the treatment of breast cancer. Extensive studies on the action of tamoxifen in various normal tissues revealed that contrary to its antiestrogen action in breast tissue, it acts as an agonist in other normal tissues such as the endometrium. This latter feature of tamoxifen has raised concerns related to the anticipated stimulation of the development of other estrogen-related cancers, such as endometrial carcinoma; however, this risk appears practically reduced as compared to the beneficial effects of tamoxifen in the management of breast cancer. It is thought that the molecular mechanism of this dual and tissue-specific action of tamoxifen is related to the fact that although the tamoxifen–ER complex still binds DNA, the conformation differs from that in the absence of tamoxifen (but with estradiol), resulting in changes in the balance of the coactivators and corepressors that also bind to the receptor and are needed to initiate transcription. Thus, since these coactivators and corepressors considerably differ both qualitatively and quantitatively in each cell type, the "net" outcome of receptor activation varies.

Since that time, various analogs of tamoxifen have been synthesized with the aim to improve its agonistic/antagonistic action in a variety of tissues and these have been termed 'selective ER modulators' (SERMs). A promising SERM is *raloxifene* that, whilst it retains its antiestrogen activity in mammary tissue (and agonistic activity in bone), has lost its estrogen-like activity in the endometrium.

In parallel with these attempts to improve the characteristics of SERMs, synthetic steroid analogs are also being developed that aim to display exclusive antiestrogen activity [selective ER downregulators (SERDs)].

GHRH and Carcinogenesis

Several different peptide and protein hormones have been associated directly or indirectly with carcinogenesis. Among them, somatostatin, a peptide hormone that regulates the production and release of GH in a negative manner, is used extensively, especially in the form of its potent agonistic analogs, for the management of various solid tumors. IGF-I is also a protein hormone (growth factor) that is an established mitogen for various cancers and is also a potential target for the development of anticancer drugs. Here we will describe in some detail the action and the potential implication in oncogenesis of a different peptide hormone, i.e. GHRH, not because of our in-depth knowledge of its role in cancer, but because it is only recently that its role in carcinogenesis has emerged, representing a promising target for the development of cancer therapeutics.

GHRH is a neuropeptide that is secreted by the hypothalamus, and is responsible for the stimulation of the synthesis and release of GH in the pituitary. However, recent evidence suggests that in addition to this action (neuroendocrine action), GHRH is also present in several extrahypothalamic tissues and is involved in a variety of cellular processes. Its function is related to the regulation of cell proliferation and differentiation of various nonpituitary cell types – importantly, in carcinogenesis. The mechanisms by which GHRH affects the peripheral, extrapituitary tissues remains poorly understood, but it is likely that it also involves paracrine and autocrine pathways in addition to the classic neuroendocrine action (Fig. 6). The action of GHRH in cancer cells is mediated, at least in part, by receptors that represent splice variants of the full-length pituitary GHRH receptor.

The study of the direct role of GHRH in carcinogenesis was initiated after the development of specific peptide antagonists that were produced almost a decade ago in an effort to downregulate GH secretion, which in turn would reduce the

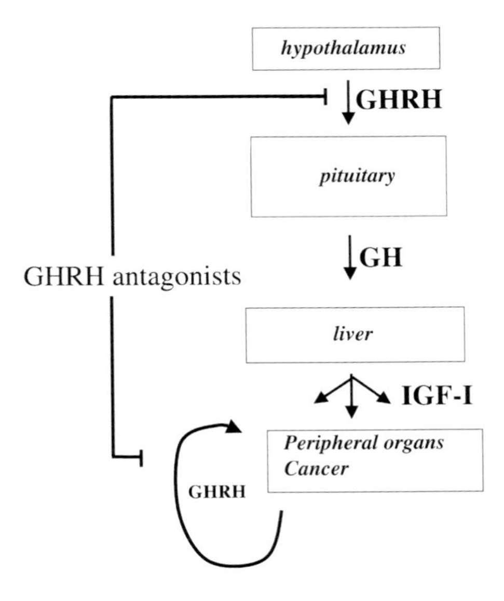

Fig. 6 Systemic and local (paracrine/autocrine) action of GHRH under normal conditions and in carcinogenesis. GHRH antagonists inhibit both the systemic and local action of GHRH.

levels of circulating IGF-I produced in the liver. IGF-I is a well-established mitogen for several cancers and reduction of its activity has been shown by various approaches to be beneficial for the management of the disease. Indeed, antagonistic analogs of GHRH, as predicted, were shown to be capable of reducing GH/IGF-I levels in experimental animals and eventually inhibiting the growth of various IGF-I-dependent neoplasms in nude mice, such as osteosarcomas, prostatic carcinoma, as well as kidney, lung and other cancers (Fig. 6). This indirect role of GHRH in carcinogenesis was also proved by experiments using "little" mice bearing a missense mutation in the GHRH receptor, demonstrating that genetic ablation of GHRH action results in reduced sensitivity to chemically induced liver carcinogenesis. Furthermore, the same animals were shown to provide a nonpermissive environment for the growth of established breast tumors, as compared to wild-type animals – inoculation of "little" mice with tumorigenic human breast cancer cells showed that the growth of the resulting xenografts was considerably compromised as compared to that of wild-type animals.

However, subsequent studies on the precise anticancer mechanism of GHRH antagonists revealed that systemic inhibition of IGF-I does not reflect the only mode of their antitumor action. Two lines of experiments showing that (i) GHRH is expressed by various primary tumors and cell lines, and (ii) GHRH and GHRH antagonists act in the opposite manner to affect the rate of cell proliferation by mechanism(s) that bypass the GH/IGF-I axis. These experiments established that this peptide hormone satisfies the criteria to be considered an autocrine/paracrine growth factor, thus suggesting that it exerts more widespread actions than those originally anticipated (Fig. 6).

Normal nonpituitary tissues that express GHRH include the placenta, ovaries, testes and lymphocytes; cancers producing GHRH include breast, endometrial, ovarian, gastric and prostatic cancers, as well as small cell lung carcinomas and malignant bone tumors. It is notable that the original isolation of GHRH was performed in pancreatic tumor extracts; however, as mentioned before, it is only recently that its role in carcinogenesis has begun to unravel.

The major obstacle in studying the extrapituitary role of this neuropeptide in carcinogenesis has been that the receptors that appear to mediate the effects of GHRH in cancers are most likely different from those that operate in the pituitary. This is due to the fact that the receptor that has been identified as the predominant (if not the exclusive) mediator of GHRH effects in the pituitary is the GHRH receptor that is a G-protein-coupled seven-transmembrane-domain receptor with a relatively narrow pattern of expression in tissues such as the kidney and the placenta, apart from the pituitary. However, recent studies showed that splice variants of GHRH receptor are widely expressed in both normal and, particularly, malignant tissues, and can mediate the mitogenic effects of GHRH. One of those splice variants, termed SV1, that differs from the original full-length GHRH receptor in a small portion of the N-terminus (encoding for a part of the extracellular domain of the receptor), is of particular importance because it possesses ligand-independent activity. Thus, it is able to stimulate cell proliferation in the absence of ligand binding. However, if GHRH ligand is present its effects are intensified (see Fig. 7).

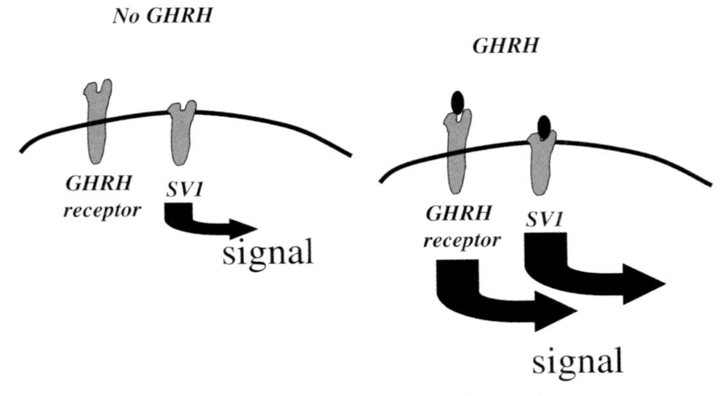

Fig. 7 Contrary to the pituitary GHRH receptor that produces only ligand-dependent activity, SV1 apparently also possesses ligand-independent activity that is intensified in the presence of its ligand.

GHRH analogs represent a class of promising anticancer drugs because they are characterized by two important advantages: (i) they are virtually nontoxic because the harmful consequences of GHRH ablation are minor in the adult organism and (ii) they are short peptides that can easily be chemically synthesized, permitting their rapid development and subsequent evaluation.

Bibliography

Castagnetta L, Granata OM, Cocciadiferro L, Saetta A, Polito L, Bronte G, Rizzo S, Campisi I, Agostara B, Carruba G. Sex steroids, carcinogenesis, and cancer progression. *Ann NY Acad Sci* **2004**, *1028*, 233–246.

Culig Z, Comuzzi B, Steiner H, Bartsch G, Hobisch A. Expression and function of androgen receptor coactivators in prostate cancer. *J Steroid Biochem Mol Biol* **2004**, *92*, 265–271.

Foster JS, Henley DC, Ahamed S, Wimalasena J. Estrogens and cell-cycle regulation in breast cancer. *Trends Endocrinol Metabol* **2001**, *12*, 320–327.

Freeman MR, Cinar B, Lu ML. Membrane rafts as potential sites of nongenomic hormonal signaling in prostate cancer. *Trends Endocrinol Metabol* **2001**, *16*, 273–279.

Johnston SD. Endocrinology and hormone therapy in breast cancer selective estrogen receptor modulators and downregulators for breast cancer – have they lost their way? *Breast Cancer Res* **2005**, *7*, 119–130.

Kiaris H, Schally AV, Kalofoutis A. Extra-pituitary effects of growth hormone-releasing hormone. *Vitam Horm* **2004**, *70*, 1–24.

Linja MJ, Visakorpi T. Alterations of androgen receptor in prostate cancer. *J Steroid Biochem Mol Biol* **2004**, *92*, 255–264.

Soronen P, Laiti M, Torn S, Harkonen P, Patrikainen L, Li Y, Pulkka A, Kurkela R, Herrala A, Kaija H, Isomaa V, Vihko P. Sex steroid hormone metabolism and prostate cancer. *J Steroid Biochem Mol Biol* **2004**, *92*, 281–286.

Swinnen JV, Heemers H, Van de Sande T, Schrijver E, Brusselmans K, Heyns W, Verhoeven G. Androgens, lipogenesis and prostate cancer. *J Steroid Biochem Mol Biol* **2004**, *92*, 273–279.

16
Viral Oncogenesis

It is estimated that viral oncogenesis accounts for about 15% of cancer cases worldwide. However, its impact in cancer biology is far more important than the percentage of cancer cases that can be associated to a viral infection in a causative manner. The study of the mechanism of action of tumor viruses has shed light on basic aspects of tumor biology and, especially during the early days of cancer research, the field (as well as the general field of molecular biology) has considerably benefited from the recruitment of many virologists into the study of cancer.

It has been proposed that four criteria should be fulfilled in order to consider a virus as oncogenic in humans:

(i) The virus (or its nucleic acid) should be present in the tumor cells.
(ii) Transfection of a certain portion of the viral genome should cause either immortalization or transformation.
(iii) Demonstration that there is a causative association between the acquisition of the malignant phenotype and (portions of) the viral genome,
(iv) Clinical relevance in terms of correlation between viral infection and increased tumor incidence.

Various different viruses have been linked to neoplasia in a causative manner, fulfilling the abovementioned criteria either fully or partly. Certain viruses belong to the group of retroviruses and contain a genome consisting of a single-stranded RNA molecule, whereas others have a genome consisting of double-stranded DNA. Their mechanism of action is quite heterogeneous. In some cases they encode for certain genes that exhibit oncogenic potential, having the ability to interfere with cellular signaling pathways and initiate malignant conversion. Examples of this case are the T antigen of viruses such as the papillomaviruses that have the ability to bind and inactivate cellular tumor suppressors exemplified by p53 and Rb. Thus, functionally they mimic a loss of (tumor suppressor) function phenotype. In other cases they induce malignancy by activating endogenous oncogenes such as *wnt* and *Notch* oncogenes by insertional mutagenesis. Indeed, following the elucidation of the oncogenic mechanism for the corresponding viruses, experimental strategies have been devised and successfully performed to identify novel cancer-related genes.

Understanding Carcinogenesis. Hippokratis Kiaris

ISBN 3-527-31486-5

Tumors that have been initiated by oncogenic viruses, not surprisingly, exhibit all of the hallmarks of malignancy triggered by virus-independent mechanisms, including genomic instability, resistance to apoptosis and increased proliferative capacity.

We will now discuss in some detail some cancer-causing viruses, with an emphasis on the mechanism by which their oncogenic potential is manifested.

Human Papillomavirus (HPV)

HPV infection has been associated with various proliferative conditions, with the most malignant being the development of cervical as well as other types of cancer (such as that of vulva, penis, anus, etc.). A long latency period is needed from the time of infection until the onset of malignancy, suggesting that accumulation of mutations in cellular genes is required along with the expression of oncoproteins encoded by HPV in order to cause malignant conversion. Furthermore, only a certain percentage of carriers develop the disease, implying differential susceptibility in the hosts for the development of the disease and/or cooperation with environmental and probably other factors.

More than 200 different HPV strains that have been identified that can be classified into two major categories – mucosal and cutaneous HPVs. Depending on their association with malignancy they are either of high or low risk, with the latter containing the vast majority of the various HPV strains. Among the high risk viruses are HPV-16, -18 and -31, which are the more oncogenically potent, and their presence displays the highest association with the development of some types of cancer.

The life cycle of HPVs is associated with the differentiation state of the infected epithelial cells. Initially, infection occurs at the actively dividing basal epithelial cells and the virus remains at low copy number during the differentiation of the infected cells. However, when these cells undergo terminal differentiation, the virus enters a high copy number "state" and progeny viruses are produced.

HPVs are nonenveloped viruses and their genome consists of double-stranded DNA that is about 8 kb long. Their genome contains three distinct regions: the early region (E) that is about 4 kb long and encodes for nonstructural proteins, the late region (L) that encodes for two capsid proteins and is about 3 kb, and the 1-kb long control region (LCR) or upstream regulatory region (URR) that contains various regulatory elements. In total, eight open reading frames (ORFs) are contained within the HPV genome that are transcribed from a single strand as polycistronic messages.

Among the virus products of the HPV genome, E1 and E2 are involved in the replication of the viral genome and bind to regions around the origin of replication. E2 is also implicated in the regulation of virus copy number.

The oncogenic properties of HPVs are mostly attributed to the products of E6/E7 genes. Expression of these proteins in tissue culture experiments was shown to cause immortalization of keratinocytes, increase the mutational rate by inducing genomic instability, and synergize with other oncogenes such as *ras* and *myc* to elicit malignant transformation. At the molecular level these effects of E6 and E7 oncoproteins are mediated by their ability to bind and inactivate the endogenous p53 and Rb tumor-suppressor genes (TSGs) that are important regulators of the cell cycle. In addition, E6 can induce the transcription of hTERT, which encodes for the telomerase reverse transcriptase, the ribonucleoprotein with an essential role in the maintenance of telomere length that is also causatively associated with cellular immortalization. Thus, the presence of E6 and E7 results in the concomitant inactivation of at least two of the cell's tumor suppressors and the induction of telomerase activity. E7 is a relatively small protein of about 100 amino acids with no intrinsic enzymatic activity. It has been shown that E7 binding affinity to Rb is higher in high-risk viruses than that in low-risk HPV viruses. Binding of E7 to Rb prevents the latter from binding to E2F, thus promoting cell cycle progression. In addition to E2F, E7 also binds to p107 and p130, preventing E2F transcription.

E6 is slightly larger than E7, consisting of about 150 amino acids, and is responsible for the ubiquitin-dependent degradation of p53. In addition, E6 can act by p53-independent mechanism(s) including those involving the interaction with various PDZ domain-containing proteins that mediate protein–protein interaction, especially in cases involving the regulation of cell–cell contact. (The PDZ domain is a conserved motif of 80–90 amino acids first identified in post-synaptic density protein PSD95, *Drosophila* TSG *dlgA*, ZO-1 tight junction protein.) The importance of this mode of action of E6 has been demonstrated in experiments involving the mutagenesis of the PDZ-binding domain of E6 that reduced its transforming activity.

As well as E6 and E7, some transforming activity is also produced by E5. It is notable that E5 represents the major bovine papillomavirus oncoprotein; however, the function of E5 remains largely unknown in HPVs.

Epstein–Barr Virus (EBV)

EBV has been associated with a variety of malignancies such as Burkitt's lymphoma, Hodgkin's disease, some T cell lymphomas, undifferentiated nasopharengyal carcinoma, gastric carcinoma, etc. Apart from these EBV-associated tumors, other malignancies such as a subset of breast and hepatocellular carcinoma may also be due, to some extent, to EBV infection.

The genome of EBV consists of linear double-stranded DNA that is about 172 kb long. EBV exhibits a tropism for B cells. EBV infection of B cells is mediated by interaction of the CD21 cellular receptor with the viral glycoprotein gp350 and the human leukocyte antigen (HLA) class II molecules that interact with viral gp42. As expected from the wide range of tumors associated with EBV infection, epithelial cells can also be infected by EBV; however, the mechanism by which this infection occurs remains incompletely understood.

A large amount of our knowledge regarding the mechanism of action of EBV has been derived form lymphoblastoid cell lines (LCLs) that correspond to transformed cell lines established from the *in vitro* culture of peripheral blood lymphocytes isolated from EBV carriers. It has been established that the transforming properties of EBV are predominantly due to the nuclear antigen EBNA-2 and EBV-encoded membrane protein (LMP)-1 viral proteins, while a critical role for EBNA-LP, EBNA-3A and EBNA-3C has also been demonstrated.

The role of EBNA-2, as well as of other viral products, in cellular transformation has been demonstrated by experiments involving recombinant viruses that had specific viral genes deleted. By performing such experiments it has been shown that introduction of the EBNA-2 gene in viruses that had it deleted restored its ability to transform B cells *in vitro*. The transforming properties of EBNA-2 are due, at least in part, to its ability to induce the transcription of various target genes by a RBP-Jκ-dependent manner, by binding to promoters containing the consensus GTGGGAA sequence. RBP-Jκ/CBF-1 is a transcriptional regulator that is also activated by Notch receptor. Notch signaling activation induces malignancies by binding to RBP-Jκ in various tissues, such as lymphomas and mammary carcinomas. In addition, EBNA2 activates the c-*myc* oncogene.

LMP-1 is the major transforming protein of EBV. It is a 63-kDa membrane protein that exhibits similarity to CD40 (a member of the tumor necrosis factor receptor family). Overexpression of LMP-1 produces diverged effects that usually characterize the action of other cellular oncogenes. These effects include the induction of antiapoptotic proteins such as Bcl-2, the production of cytokines such as interleukin (IL)-6 and -8, the activation of phosphatidylinositol-3-kinase, etc.

Hepatitis B Virus (HBV)

HBV is also a double-stranded DNA virus that is associated with a significant subset of hepatocellular carcinomas. Among the various products of the viral genome, *HBx* is thought to play the major role in cellular transformation. *In vitro* experiments have demonstrated that *HBx* overexpression can transform immortalized rodent cells, while experiments with transgenic mice showed that it synergizes with other oncogenic stimuli, such as those elicited by other oncogenes or chemical carcinogens, to facilitate malignant transformation. However, although the precise role of *HBx* in carcinogenesis remains unclear, it is thought that most likely it does not behave as an oncogene *per se* but rather cooperates with other oncogenes to accelerate malignant transformation.

BK Virus (BKV)

The role of BKV in carcinogenesis is still under debate because its genome is present in a large subset of the population and can be detected in many normal tissues without pathogenic consequences. However a series of experiments have demonstrated that it possesses oncogenic potential.

BKV is a polyomavirus and like other polyomaviruses is renotropic (the kidney is the primary site of latency in humans); it can also be detected in other sites such as the stomach, lungs and lymph nodes. The oncogenic action of BKV is predominantly due to the properties of two oncoproteins encoded by its early region: large T antigen (*Tag*) and small t antigen (*tag*). *Tag* has the ability to bind and inactivate p53 and Rb tumor suppressors in a similar manner to that described for the E6/E7 oncoproteins of HPV. Thus, since these important tumor suppressors need to be inactivated constantly to maintain the malignant phenotype, consistent expression of *Tag* is needed in tumors. Furthermore, in addition to this mechanism, *Tag* also induces genomic instability as exemplified by the accumulation of chromosomal gains, deletions, gaps, breaks and other chromosomal alterations. Small t of BKV (*tag*) is thought to synergize with *Tag* in malignant transformation.

In vitro, in tissue culture experiments, *Tag* was shown able to transform rodent cells, cooperating with another oncogene such as *ras*. In humans, this effect of *Tag* was not that striking, but this should not be surprising since transformation of human cells is generally considered less efficient than that of rodent cells. *Tag*'s oncogenic potential has also been demonstrated in transgenic mice expressing *Tag*. These animals developed hepatocellular carcinoma, renal tumors and lymphoproliferative disease.

However, while the oncogenic properties of *Tag* (and *tag*) have been shown experimentally *in vitro* and *in vivo*, data from primary human tumors has cast some doubt on the potential causative role of BKV in human carcinogenesis. Indeed, according to published data (reviewed by Tognon et al., 2003), the prevalence of BKV varies considerably even between tumors of the same type. Furthermore, the virus load is quite small, estimated to less than one virus copy per cell, while strong heterogeneity in infection exists at the cellular level in a given BKV-positive tumor with not all cancer cells carrying the viral genome. The hypotheses that have been proposed to overcome these arguments involve the "hit and run" mechanism according to which the initiation of genomic instability and the subsequent oncogenic changes are triggered by the virus, which is not then needed during the subsequent steps of the carcinogenic process. In addition, a paracrine mode of action has been proposed for BKV, according to which a single positive (infected) cell undergoes changes that can affect adjacent, BKV-negative cells. More studies are needed to elucidate the precise role of BKV in carcinogenesis.

Human T Cell Leukemia Virus (HTLV)

Until now we have described tumor viruses that contain DNA as genetic material. However, HTLV-1, the virus that is associated with adult T cell leukemia, is a retrovirus, i.e. its genetic material is RNA that during specific stages of the virus's life cycle is reverse transcribed into DNA.

HTLV-1 provirus has the typical genomic structure of other retroviruses: it contains the *gag*, *pol* and *env* genes flanked by long terminal repeats (LTRs). In addition, it contains additional genes including the *Tax* gene that plays an important role in regulating virus gene transcription, replication and proliferation in infected cells. *Tax* is considered the major transforming protein of HTLV-1 as demonstrated by experiments *in vitro* and in transgenic mice. The effects of *Tax* are mediated through interaction with the cAMP responsive element binding protein (CREB), SRF, NF-κB and AP-1, and result in the induction of cellular cytokines and antiapoptotic genes such as *Bcl-*x_L and *survivin*. In addition, *Tax* interferes with the action of p53 and p16, inhibiting apoptosis and inducing genomic instability. The precise mechanism of these interactions remains obscure; however, in certain cases it is thought that it may involve p300/CBP or NF-κB. Inhibition of signaling triggered by transforming growth factor-β, a cytokine that has a negative role in cell proliferation, has also been reported in HTLV-1-infected cells and is mediated by *Tax*. Finally, *Tax* protein of HTLV-1 contains binding motifs for the PDZ domain proteins (see also E6 protein of HPV above) with their apparent consequences in mediating cell–cell interactions during tumorigenesis. The contribution of these domains to the oncogenic properties of HTLV-I *Tax* protein becomes apparent from the fact that *Tax* protein of HTLV-II, a similar retrovirus with no oncogenic potential, differs only in the absence of these domains. It is noteworthy when *Tax2* (*Tax* of HTLV-II) is genetically engineered to contain these domains from *Tax1*, it acquires oncogenic potential.

The precise details of the mechanism of HTLV-I infection remain under investigation. However, it is thought that HTLV-I virions are transmitted through "cell–cell" contacts from the infected into the target cell. The glucose transporter 1 has been recognized as a major receptor mediating the transmission of HTLV-I. Following introduction into the target cells, the viral genome integrates randomly into the host's genome and *Tax* protein induces the proliferation of the infected cells. Subsequently, *Tax* expression must be downregulated because it is thought that *Tax*-expressing cells constitute targets for cytotoxic T cell lymphocytes. Various mechanisms have been proposed to account for the suppression of *Tax* expression following infection. These mechanisms include hypermethylation of the 5′-LTRs, mutational inactivation of the *Tax* gene, etc. Furthermore, p30, *Rex* and HBZ viral accessory proteins also suppress *Tax* expression.

A latency period of a few decades is usually required from the time of infection until the onset of malignancy, suggesting that the accumulation of several genetic hits is needed to cooperate with *Tax* oncoprotein. Among these hits, inactivation of p53, silencing of p16, and alterations in $p27^{kip1}$ and Rb have been detected in adult T cell leukemia/lymphoma patients. However, the precise mechanism of HTLV-I-mediated oncogenesis remains to be elucidated.

Mouse Mammary Tumor Virus (MMTV)

Despite the fact that MMTV infects, and thus causes cancer in, mice, several arguments justify its inclusion in the description of various tumor-associated viruses. The first such argument is related to the fact that the mechanism of MMTV-induced oncogenesis is distinct from the mechanisms described above for other viruses that have been associated with human neoplasia. This mechanism involves insertional mutagenesis instead of expression of specific oncoproteins with defined properties. Second, investigations employing MMTV infections as a model have contributed a lot to the understanding of the basic aspects of multistage carcinogenesis, including the identification of novel oncogenes important for human disease. Furthermore, the regulatory sequences of MMTV, i.e. LTRs, provided the means to develop "onco-mice" – transgenic mice that develop mammary cancer due to the overexpression of specific oncogenes. Finally, recent reports provide evidence that MMTV-related sequences have been identified in primary human breast cancers as well, suggesting that this (or a closely related virus) may be responsible for a subset of human breast cancers.

The isolation and identification of MMTV, and the demonstration of its causative relationship with breast cancer, followed the development of breast cancer-prone strains of mice about 70 years ago, such as C3H, BR6, etc. Studies using these animals showed that the pattern of inheritance of breast cancer susceptibility does not conform to a Mendelian mode of inheritance, but rather implies some "extrachromosomal" inheritance associated with the development of breast cancer in the mother. Subsequent studies demonstrated that, indeed, an infectious agent is responsible for the development of the disease and that it is transmitted through the milk.

The mechanism by which MMTV induces carcinogenesis is related to the integration of the MMTV genome into the cell's genome that results in the activation of endogenous cellular genes under the regulation of the MMTV LTRs that operate as promoters/enhancers. Analyses of the cellular genes that become activated following MMTV infection show that most of the time the virus exhibits some specificity against the cellular targets, being integrated at sites that activate a limited repertoire of endogenous genes. Those include the *wnt* genes that are members of a family of more than 12 genes homologous to the fly gene *wingless* involved in the regulation of segment polarity. Members of the fibroblast growth factor family are also activated following MMTV infection, as well as Notch receptor genes that are involved in the regulation of cell fate decisions. In all cases viral genome integration results in the overexpression of the endogenous genes, either intact, as in *wnt* and *fgfs*, or truncated, as in Notch receptors, resulting in the production of a form of a protein that functionally mimics the activation of the Notch receptor(s) by its ligand(s). Importantly, in all cases it has been demonstrated that transgenic mice that express these oncogenes in the mammary epithelium, under the regulation of MMTV LTRs, develop breast cancer.

As already mentioned above, MMTV was thought to infect only mice and was associated only with mouse breast cancer. However, relatively recent studies have shown that some human breast cancers bear sequences that are homologous to the MMTV *env* gene, implying the viral origin for a portion of these tumors. Indeed, subsequent studies have confirmed these initial findings and extended the observation to the identification of sequences corresponding to the whole MMTV genome [now termed human mammary tumor virus (HMTV)] and its absence from normal tissues of infected individuals. The impact of these findings also led to the suggestion of the intriguing hypotheses of the trans-species infection of MMTV that is responsible for breast cancer in women. These hypotheses are based on the fact that in geographic areas where *Mus domesticus* is the predominant mouse species, with a high incidence of MMTV, breast cancer is common, whereas it is not in areas in which the predominant mouse species are *Mus musculus* and *Mus castaneus,* with a low MMTV content. Various considerations arise from these findings before we reach the point where we can etiologically associate these sequences with human breast cancer. Importantly, demonstration by functional assays that these sequences confer oncogenic potential has yet to be shown.

Bibliography

Callahan R, Smith GH. MMTV-induced mammary tumorigenesis: gene discovery, progression to malignancy and cellular pathways. *Oncogene* **2000**, *19,* 992–1001.

Fehrmann F, Laimins LA. Human papillomaviruses: targeting differentiating epithelial cells for malignant transformation. *Oncogene* **2003**, *22,* 5201–5207.

Gatza ML, Chandhasin C, Ducu RI, Marriott SJ. Impact of transforming viruses on cellular mutagenesis, genome stability, and cellular transformation. *Environ Mol Mutagen* **2005**, *45,* 304–325.

Holland JF, Pogo BGT. Mouse mammary tumor virus-like viral infection and human breast cancer. *Clin Cancer Res* **2004**, *10,* 5647–5649.

Matsuoka M. Human T-cell leukemia virus type I (HTLV-I) infection and the onset of adult T-cell leukemia (ATL). *Retrovirology* **2005**, *2,* 27.

Munger K, Baldwin A, Edwards KM, Hayakawa H, Nguyen CL, Owens M, Grace M, Huh KW. Mechanisms of human papillomavirus-induced oncogenesis. *J Virol* **2004**, *78,* 11451–11460.

Tognon M, Corallini A, Martini F, Negrini M, Barbanti-Brodano G. Oncogenic transformation by BK virus and association with human tumors. *Oncogene* **2003**, *22,* 5192–5200.

Young LS, Murray PG. Epstein–Barr virus and oncogenesis: from latent genes to tumours. *Oncogene* **2003**, *22,* 5108–5121.

Young LS, Rickinson AB. Epstein–Barr virus: 40 years on. *Nat Rev Cancer* **2004**, *4,* 757–768.

Part IV
Unifying the Concepts

Understanding Carcinogenesis. Hippokratis Kiaris

ISBN 3-527-31486-5

17
Cooperation of Multiple Biological Processes is Needed for the Development of Fully Fledged Malignancy

Previously, we have described classes of individual genes that are etiologically implicated in the progression of cancer. However, we have also mentioned that practically all of them govern more than one distinct processes, exhibiting a pleiotropy in their function. Despite this heterogeneity in the function and types of alterations of cancer-related genes, some essential processes in cellular physiology must be altered for cancer progression. Whilst these processes can be induced by alternative pathways in different cancers, the biological changes they dictate are common in different lesions, regardless of their origin. Each of these processes that must be altered during carcinogenesis represents an important point in the cell's defense mechanisms that potentially may stop the growth of cancer. An outline of these processes is given below.

Initially, a cell that is going to develop clonally into a malignant tumor must have the balance between cell division and cell death shifted towards the direction of cell proliferation. In order to do so it must sensitize its perception to growth factors and/or develop new autocrine stimulatory loops. By doing this it is arrested in a proliferative state. At the same time, resistance against the cell death-inducing pathways must be acquired or apoptosis will inhibit the clonal expansion of the overproliferating cells. However, entry into a highly proliferative state is not sufficient for continuous growth because, beyond a certain number of doublings, normal cells undergo replicative senescence. Therefore, mechanisms bypassing this safeguarding state should be devised. In most of cases, this unlimited replication capacity is afforded by the reactivation of telomerase activity, which is responsible for the immortal phenotype of the cancer cells.

The abovementioned changes predominantly correspond to cell-autonomous processes, i.e. the same cells that are destined to become malignant are the targets of these alterations. However, growing lesions should also interact with their microenvironment, and thus need to be subject to additional changes that permit efficient crosstalk with different cell types and regulate events that are essentially noncell autonomous. We have already described how interruption of these heterotypic signaling cues is sufficient to inhibit tumorigenesis. At this stage stromal fibroblasts are engaged in the carcinogenic process and while at first they most likely protect the organism from the developing tumor, they subsequently acquire cancer-associated characteristics and become cancer-associated

Understanding Carcinogenesis. Hippokratis Kiaris

ISBN 3-527-31486-5

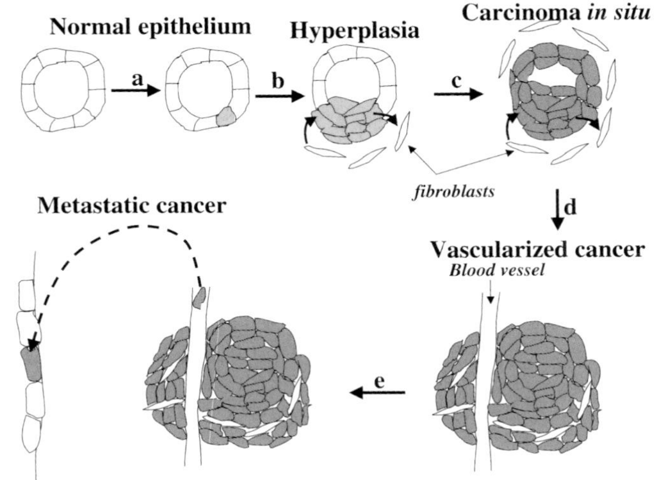

Fig. 1 Typical physical history of carcinogenesis in a hypothetical glandular epithelium. (a) Specific mutations or overproduction of growth factors initially render a cell (or cells) susceptible to overproliferation, finally resulting in the development of hyperplasia that is initially typical and subsequently atypical (b). Stromal fibroblasts at this stage initially play a negative role, although subsequently they play a positive role in the carcinogenic process. Accumulation of additional mutations results in the development of a carcinoma *in situ* (c). This histopathological entity, although characterized by the presence of cancer cells, it is not yet considerably vascularized, the overall tissue architecture is not altered and, importantly, is by definition not invasive. Further growth of this malignant lesion triggers angiogenic pathways (d) and, finally, the tumor becomes metastatic (e). Light and dark grey indicates precancerous and cancer cells respectively.

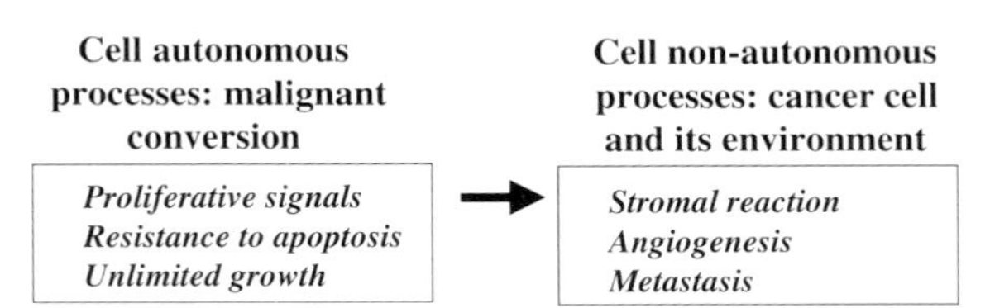

Fig. 2 Molecular alterations in carcinogenesis can be classified as those related to the malignant transformation (cell autonomous) and those associated with the interaction of a cell with its microenvironment (cell non-autonomous). The first type of alterations usually must occur before the second type.

fibroblasts. Following this transition into the cancer-associated state they play a positive role in carcinogenesis as they provide the (cancer) cells with growth factors. Subsequently, the increasing needs of tumors for oxygen and nutrients dictate the operation of an angiogenic switch that triggers tumor neoangiogenesis and the corresponding lesions become highly vascularized. Finally, cancer cells metastasize into sites of secondary growth and the tumor becomes metastatic (Figs. 1 and 2).

This sequence of events described above, which varies depending on the exact type of cancer evaluated, is applicable to practically each and every malignant tumor.

Bibliography

Hahn WC, Weinberg RA. Modeling the molecular circuity of cancer. *Nat Rev Cancer* **2002**, *2*, 331–341.

Hanahan D, Weinberg RA. The hallmarks of cancer [Review]. *Cell*, **2000**, *100*, 57–70.

18
Carcinogenesis *In Vivo*: Animal Models and Basic Approaches to Generate Genetically Modified Animals

Advances in molecular biology during the last two decades have offered the possibility to use experimental animals (particularly mice) as cancer models. Despite certain limitations that such models display for the study of human cancer, the information they provide and their contribution to understanding fundamental aspects of the disease in man is incomparable to that obtained by any other system used to date.

The use of animals in research, and the development of appropriate methodologies to generate and manipulate them, has preoccupied scientists for centuries. Van Dyke and Jacks (2002) mentioned that in the 17th century making a mouse was thought to be quite easy – all you needed, according to the famous alchemist/chemist Jean Babtista Van Helmont, was to "stuff a sweaty shirt into a flask of wheat and incubate for 3 weeks". It is obvious that this notion was soon proven wrong, reaching a point today where "mouse making" has been advanced enormously

Several categories of genetically engineered animals have been developed, differing in the type of the modification and the methodology followed, and in the kind of information they can offer. Genetically engineered mice are used in virtually all fields of biomedical research, particularly in experimental/molecular oncology in which they have been very informative in both basic research and preclinical studies. Depending on the type of the genetic alteration that they have been engineered to harbor, the function of one or more tissues may be affected, they may involve the introduction and subsequent expression of new alleles of disease-related genes, they may have specific endogenous genes suppressed or they may express specific genes only when and where the researcher dictates.

An account of the main principles governing the development and use of genetically engineered mouse models of carcinogenesis will follow in this chapter, accompanied by a description of some such models and their contribution to molecular oncology.

Understanding Carcinogenesis. Hippokratis Kiaris

ISBN 3-527-31486-5

Why Genetically Engineered Mice?

Cancer is a multistage disease. Consistent with this notion, the malignant transformation of a normal cell must involve the accumulation of several mutations in order to acquire and exhibit the phenotypic characteristics of a cancer cell. As became clear in previous chapters, following malignant transformation and tumor progression cancer cells continue to accumulate mutations, at a rate that surpasses that of normal cells, resulting in the acquisition of mutations continuously affecting their characteristics. These new or modified phenotypic properties may be reflected in their metastatic and invasive properties, resistance to anticancer therapy, proliferation rate, and virtually every other aspect constituting the malignant phenotype.

During earlier years of research, experimental studies were thought to be informative even if they were performed exclusively in cells cultured *in vitro*. However, it soon became apparent that several properties of the cancer cells, specifically those related to the bilateral interaction of the cancer cells with the surrounding normal tissue, could not be seen in cells in the culture dish. Importantly, the latter becomes apparent when considering recent views that consider the tumor not as a homogenous mixture of cancer cells that behave in an autonomous manner, but rather as an heterologous system of different types of cells, both malignant as well as genotypically normal that interact with, as well as being affected, by their environment (see Chapter 7).

Tumor Transplantation Experiments: Xenograft Development

The aforementioned disadvantages have been resolved (or partially resolved) by devising experiments performed in immunocompromised animals – most importantly in mice that could be inoculated with cancer cells, usually subcutaneously, and subsequently grow solid tumors that occasionally metastasize. Using more delicate surgical procedures the cancer cells could be implanted orthotopically, i.e. in the same tissue from which the tumor had been derived. For example, inoculation in the mammary fat pad of breast cancer cells or intracranial implantation in the case of brain cancers results in the development of tumors that simulate with even higher accuracy the primary tumors that develop in humans. Indeed, this ability of cancer cells to grow tumors in immunocompromised mice, i.e. the tumorigenicity, constitutes a landmark phenotypic manifestation of the malignant phenotype and an essential diagnostic test to classify a particular cell as being subjected to malignant transformation. While very informative in certain aspects of oncological studies, especially with regard to the assessment of the efficacy of novel anticancer drugs, the aforementioned approaches show a crucial limitation – they only allow the study of a tumor *after* it has been developed and not *during* malignant transformation. In other words, the resulting tumor is not primary, but in the wider sense it is by definition metastatic.

Chemical and Viral Carcinogenesis in Mice

This last disadvantage of experiments performed by inoculating immunocompromised mice with cancer cells has been solved by performing experiments in which mice are administered with specific chemical agents, apparently mutagens, or infected with oncogenic viruses, forming tumors that are indeed primary. Selection of an appropriate oncogenic stimulus, a mutagen or a virus, can induce cancer formation by introducing quite specific mutations. For example, H-*ras* mutations are predominantly found following treatment of mouse skin with specific carcinogens such as *N*-methyl-*N'*-nitro-*N*-nitrosoguanidine (MNNG), methylnitrosourea (MNU) and 7,12-dimethylbenz[*a*]anthracene (DMBA). Alternatively, insertional activation of the *wnt* oncogenes in breast cancer occurs almost deterministically following infection with the mouse mammary tumor virus (MMTV). Again, strictly mechanistic conclusions regarding carcinogenesis in these models cannot be obtained, since they formally represent molecular epidemiological studies. The latter is apparent when considering the possibility of yet unidentified mutations in these tumoral lesions, apart from the ones we consider important. However, in certain cases the correlation between tumorigenesis and the identification of specific genetic alterations reaches very high statistical significance that permits causative conclusions to be drawn.

Therefore, during the last 20 years the technology for the introduction of specific mutations in experimental mice has been advanced to such a high level that it allows the manipulation of the mouse genome in a manner that not only the tissue, but also the timing at which the genetic alteration is induced can be selected. Apparently, the latter permits the determination, in a causative manner, of the association between specific genetic alterations and particular phenotypic characteristics, diminishing the statistical parameter to a negligible level.

Strategies for Introducing Mutations in the Genetic Material – Major Types of Genetically Engineered Animals

Transgenic Mice

The most simple and historically first approach resulting in the specific modulation of the genetic material of experimental animals is transgenesis. This strategy consists of the introduction of a gene, usually in the form of cDNA or sometimes a minigene containing one or more introns, in the gametic line of a mouse. This modification results in the overexpression of the gene or cDNA of interest in the tissue(s) dictated by the promoter that has been selected. This method is the approach of choice for the study of oncogenes that, as already mentioned, are usually overexpressed in primary tumors. It has to be noted that application of this approach in the study of tumor-suppressor genes (TSGs), besides the fact that it does not simulate the *in vivo* conditions and therefore it would not be very informative (tumor suppressors almost exclusively become inactive in tumors), it is

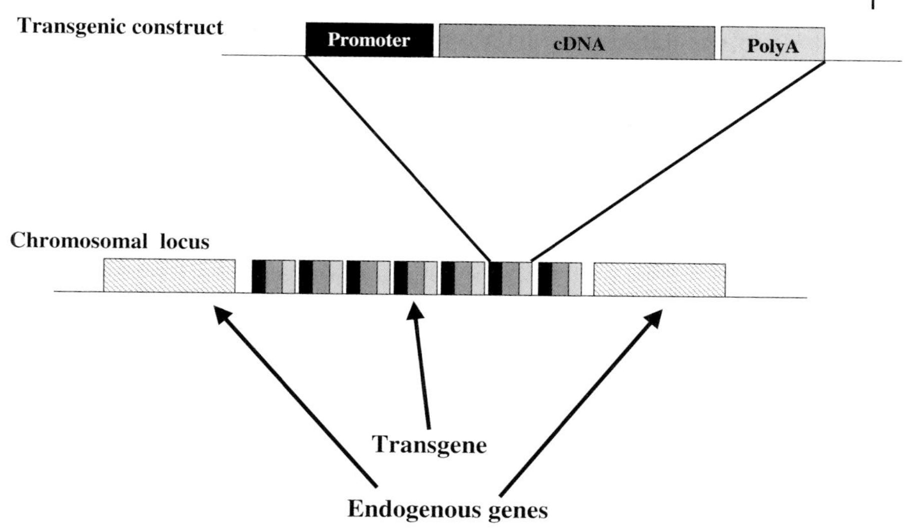

Fig. 1 Diagrammatic illustration of a typical construct used to generate transgenic animals. The cDNA of interest is cloned between a promoter, which can be either constitutively expressed in all tissues or tissue specific, and a polyA signal. After restriction digestion and removal of the unnecessary sequences, the DNA containing the cDNA of interest is microinjected into fertilized oocytes and integrates randomly in the genome as a multimer. Finally, the offspring are tested for the presence of the transgene by the polymerase chain reaction and/or Southern blot analysis.

also not going to be experimentally feasible. This is a consequence of the fact that TSGs function in a manner consistent with the inhibition of cell proliferation and/or induction of apoptosis, and therefore, their overexpression would result in lethality or developmental aberrations in the tissues in which this gene is expressed. A diagrammatic representation of an "artificial gene" (construct) used for such analyses is shown in the Fig. 1. The cDNA that is under investigation is cloned between a polyadenylation (polyA) signal and a specific promoter, selected on the basis of the specific expression pattern desired. Subsequently, after endonuclease digestion to remove the sequences unnecessary for transgenesis, it is introduced by microinjection into the fertilized oocytes (usually) where it integrates at random sites into their chromosomes. When the zygote reaches the two-cell embryo stage it is implanted in a pseudopregnant foster mother.

The random integration sites of the transgenes into the chromosomes of the host, in association with their multiple copy number (by the "head to tail" pattern), results in some, usually considerable, variability of the phenotype in the transgenic animals. Therefore, in order to avoid confusion due to the inactivation (insertional mutagenesis) of endogenous genes and to obtain reliable information, it is mandatory to obtain and study in parallel multiple independent transgenic lines with the same transgene incorporated.

Studying Gene Products

As already mentioned, the expression pattern of the cDNA used for the generation of transgenic mice is ideally solely determined by the promoter selected for the development of the appropriate construct. Thus, if a promoter with a global pattern of expression is used [i.e. phosphoroglycerine kinase (PGK); z-globin] the gene of interest is expressed in virtually all tissues of the animals, notwithstanding the fact that some variation in the expression levels in different tissues is unavoidable. It is apparent that in certain cases the overexpression of a specific gene, i.e. an activated oncogene, can be deleterious during embryogenesis, producing a lethal phenotype. In such cases, a more restricted pattern of expression is required for the transgenesis which can be achieved by the selection of tissue-specific or tissue-restricted promoters. Such promoters are the promoters of the keratins that, depending on the specific keratin subtype selected, drive the expression of the gene of interest in specific layers or cell subpopulations of the skin. Another type of promoter driving a tissue-specific pattern of expression for the transgene is the promoter/enhancer of MMTV, which dictates expression in the mammary epithelial cells (breast tissue) as well as in the salivary glands of the transgenic animal. In pioneering experiments using transgenic constructs based on this promoter, Leder and coworkers (Sinn et al., 1987) developed transgenic animals which expressed the activated oncogene c-*myc* in breast tissue. Breast adenocarcinomas were developing in a stochastic manner in these animals, exemplifying the power of transgenic technology in experimental cancer research. Subsequently, the same research group showed by breeding MMTV-v-H-*ras* with MMTV-c-*myc* animals that bitransgenic MMTV-v-H-*ras*/c-*myc* mice exhibited an accelerated rate of tumorigenesis, proving the synergistic action of oncogenes that cooperate *in vivo* in malignant transformation. According to the investigators, since these tumors arise stochastically and are apparently monoclonal in origin, additional somatic events appear necessary for their full malignant progression, even in the presence of activated v-Ha-*ras* and c-*myc* transgenes – underlining the multistage nature of cancer. Animals expressing oncogenes under MMTV regulation, such as v-H-*ras* and c-*myc*, also develop salivary tumors, again stochastically, in addition to the breast cancers because some (frequently considerable) activity for the MMTV promoter occurs in this particular tissue.

Tissue-specific promoters are available that can be used for the development of transgenic animals, for virtually every cell type and up to the degree that it can be differentiated by the expression of specific transcripts.

Studying Promoters

Examples like those mentioned above employ transgenic technology in order to study the consequences of the overexpression of a specific gene in a given tissue or a specific allele of a given gene in the tissue(s) that has been determined by the selected promoter. An analogous approach can also be employed in cases where it is not a specific cDNA, but rather a specific promoter that is under investigation. In this case the promoter of interest is cloned upstream, therefore dictating the expression of a protein marker that can easily be identified. Such widely used protein markers include Green Fluorescent Protein (GFP), which transmits fluorescent radiation when exposed to UV light. A protein with a similar use in transgenic technology is the enzyme β-galactosidase, which upon addition of specific substrates produces a colored (blue) product. Thus, mice generated using these experimental approaches will fluoresce when exposed to UV light or will have a blue color in the cells in which the promoter under investigation is active. Using a knock-in approach (see below) for the introduction of the reporter protein, the activation profile of the promoter of interest can be assessed with remarkable precision. Such experiments have recently been performed using the promoter of the vascular endothelial growth factor (VEGF) that drives the expression of GFP. These experiments underlined in a very elegant manner the importance of tumor–stroma interactions and the spatial, as well as temporal, pattern of expression of VEGF during tumorigenesis. The researchers that developed this system showed that implantation of cancer cells in such transgenic animals expressing GFP under the VEGF promoter results in the accumulation of green fluorescence caused by tumor induction of host VEGF promoter activity, whereas at later stages of tumor growth the fluorescent cells invade the tumor and can be seen throughout the tumor mass. However, when double-transgenic mice were generated bearing activated oncogenes along with the GFP marker, the spontaneous mammary tumors that have been elicited by oncogene expression in the VEGF-GFP mouse show strong stromal, but not tumoral, expression of GFP. It is notable that the predominant GFP-positive cells are fibroblasts and not the cancer cells in both wound and tumor models. These findings suggest that the VEGF promoter of nontransformed cells is strongly activated in a nonautonomous manner by signals elicited by the cancer cells, underlining the significance of the tumor microenvironment in carcinogenesis.

Targeted Mutations, Knocking-out/Knocking-in Specific Genes

It has already been noted that oncogenes usually behave in a dominant manner during carcinogenesis. Thus, their overexpression in transgenic mice constitutes an accurate approach to understand their function, despite the limitations of the method. However, as opposed to oncogenes, TSGs usually become inactivated by chromosomal deletions, mutations or other genetic and epigenetic mechanisms. Therefore, the generation of an animal model to study their function requires these genes to be inactivated. The latter can be achieved by the introduction of plasmids that have a structure illustrated in Fig. 2 into the embryonic stem cells of mice. In a simplified version of this strategy these plasmids must contain at least the following elements: They need to have two regions of the gene that is going to be deleted, flanking a gene that offers resistance to an antibiotic, operating as a selection marker. Following homologous recombination the positive cells

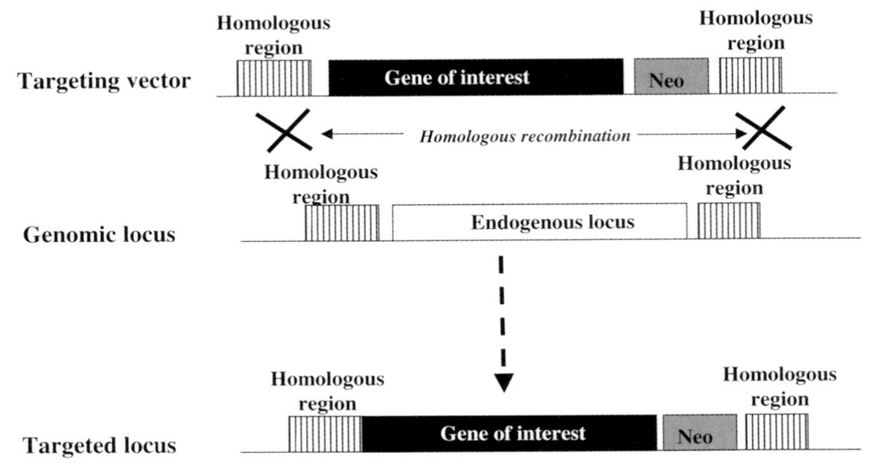

Fig. 2 Typical construct used for knocking-in or knocking-out genes. The gene of interest, which is required to be knocked into a specific locus, is cloned between homologous regions of this locus also bearing a selection marker which offers resistance to an antibiotic, usually neomycin (*Neo*). The procedure results in homologous recombination between these homologous regions that introduces the gene of interest, and the selection marker, into the targeted locus, thus inactivating it. This procedure describes the knocking-in approach. For the knocking-out procedure, similar constructs are used with the exception that only the antibiotic resistance gene is located between the homologous regions.

will have exchanged the corresponding chromosomal and plasmid regions. The end result of this process is the generation of cells that have the gene of interest interrupted by a gene that offers resistance to an antibiotic. These cells are then injected into blastocysts, which are transferred into pseudopregnant host mothers. The offspring are tested for germline transmission of the targeted locus in which one allele of the gene of interest will be deleted (knocked-out). Selective breeding of the heterozygous mice will result in the development of mice that are defective (in terms of homozygosity) for the gene that is under investigation.

Experimental animals that have the TSGs deleted exhibit increased susceptibility to carcinogenesis, as expected. A classic example is offered by the deletion of the p53 TSG, which results in increased susceptibility to lymphomatogenesis. Interestingly, even animals heterozygous for the p53 deletion are prone to the development of lymphomas by a mechanism that frequently involves the deletion of the remaining wild-type p53 allele, consistent with Knudson's "two-hit hypothesis", simulating the molecular mechanism for pathogenesis of the disease in primary human tumors. As might be expected, p53 heterozygous animals are very prone to carcinogenesis induced by chemicals, facilitating studies exploring the synergy between oncogenes and TSGs in tumor growth.

A similar approach is also used for the introduction of specific mutations in the gametic line of experimental mice. A very important characteristic of this methodology, i.e. "knock-in" technology, is that the expression of the exogenously introduced gene is regulated by endogenous regulatory elements, which apparently results in a pattern of expression that is identical to that of the endogenous gene. Technically, the generation of the knock-in animals is very similar to the strategy followed for the knocking-out of genes with the exception that the construct (plasmid) used for the recombination (see above) contains the cDNA of the gene which we want to "knock-in" flanked by the homologous regions of the gene we want to replace. This methodology, depending on the application, can be employed either in order to introduce specific mutations in a gene or to replace an endogenous gene with another, not necessarily encoded by the same genetic locus. Such an experimental approach has been performed in order to test if different cyclins, such as cyclin D1 and E, which encode for proteins with an essential role in the regulation of cell cycle progression, are functionally interchangeable. Using the aforementioned approaches, many different lines of mutant animals have been developed that have been very informative for the study of carcinogenesis. For example, knock-in mice were generated which had cyclin D1 deleted and expressed cyclin E under the regulation of the endogenous cyclin D1 promoter. It was found that the phenotype of the mutant cyclin D1 mice was completely rescued by the ectopic expression of cyclin E, suggesting that the latter can substitute for the former.

Introduction of Conditional Mutations in the Germline

Although very informative in certain cases, the methodologies mentioned above have some intrinsic limitations. In transgenic mice, apart from the integration of the transgene at random sites into the host's genome and their multiple copy numbers, some unavoidable expression frequently occurs at considerable levels in other tissues than those dictated by the specific promoter selected for the analysis. Furthermore, it is not uncommon that the specific transgene is also expressed in certain stem cell populations, resulting in conditions where the phenotypic characteristics observed are not due to the effects on cell growth/apoptosis (as usually needed for cancer-related studies), but rather to the differentiation of the cells and the acquisition of alternative cells fates.

Appropriate strategies have been developed in order to avoid such limitations through the conditional introduction and expression of the mutations under investigation. Such approaches facilitate studies in which the researcher is able to achieve temporal as well as spatial analysis of the mutations in specific tissues very accurately. Technically, the approaches that have been developed are numerous and have several modifications. Only some major strategies will be mentioned below.

Temporal–Spatial Regulation of Gene Expression using Transgenesis

This approach is shown in Fig. 3. In this case, transgenic mice are developed in which the transgene of interest is expressed not under the regulation of a promoter that is tissue specific, but rather by an "inducible" promoter that is active only in the presence of a specific transcription factor or cofactor, the expression of which can be regulated exogenously by a small molecule, e.g. a hormone. This transcription factor is selected not to be endogenously expressed (in order to avoid any crossreaction with endogenous regulatory circuits) and is produced by transgenic mice that also have to be generated using appropriate constructs, usually bearing tissue-specific promoters. In order to perform the experiment for the conditional induction of the transgene of interest, mice from the two transgenic lines, i.e. the one encoding for the transgene of interest and the other with inducible promoter, must be bred in order to obtain bitransgenic animals. The most widely used system for such studies is the tetracycline system. In this system a tissue-specific promoter drives the expression of a transactivator (TA), which in the presence of tetracycline can be activated and induce the expression of a gene (cDNA) which is under the regulation of regulatory elements containing a specific operator only in the tissues in which the TA is expressed.

Analogously, systems have also been developed for the ecdysone receptor which activates transgene expression only in the presence of the insect steroid hormone (or its chemically synthesized analog).

Similar systems, again based on tetracycline, have also been developed in order to produce exactly the opposite effect from that described above. In such systems

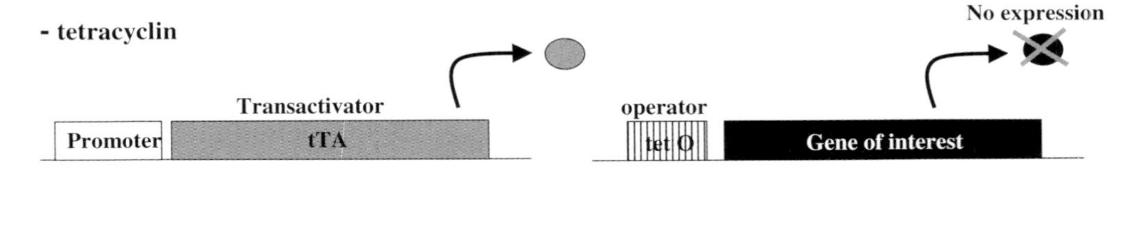

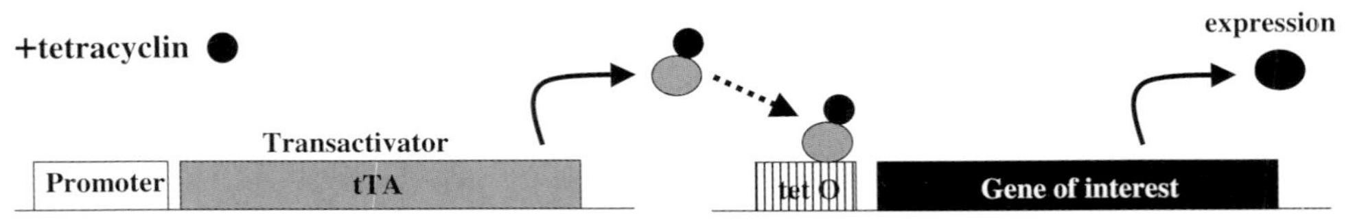

Fig. 3 Inducible transgene expression using the tetracycline (*tet*) system. A global constitutive or tissue-specific promoter drives the expression of a transactivator, while the gene of interest is under the control of an operator which requires the tetracycline-bound transactivator for activation. In the absence of tetracycline the conformation of tTA does not permit binding to the operator and, thus, the gene of interest is silenced (upper panel). When tetracycline is administered in the mice the transactivator is activated, binds to the operator and the gene of interest gets expressed (lower panel).

the addition of tetracycline results in the suppression of transgene expression. This is feasible by developing transcriptional activators that are active in the absence of tetracycline [reverse TA (rTA)], whereas in its presence they become inactive and thus silence the transgene.

The aforementioned methodologies involve the regulation of transgene expression at the level of transcription. Additional strategies involving the regulation of protein activity have also been developed. The most commonly used is the generation of chimeric proteins in which the gene of interest is fused to a segment encoding a ligand-regulated domain, such as the estrogen receptor ligand-binding domain.

A major advantage of these conditional expression-based approaches is the reversibility of transgenic expression. These systems have been used very efficiently to test the requirement for continuous expression of specific oncogenes for tumor maintenance. Experiments based on this system have been performed with various oncogenes, such as activated H-*ras*, *myc*, *neu*, etc., showing that as soon as oncogene expression was suppressed, the tumor (at least in the majority of the cases) regressed. Such experiments, apart from their significance in understanding basic aspects of tumor biology, also provide the rationale for the blockade of specific oncogenic stimuli as a means of antitumor therapy.

An additional advantage of these systems is that by generating animals bearing the transgene of interest (i.e. oncogene), many strains of mice that harbor the transactivator transgenes are available that can be crossed with each other to facilitate a wide range of such studies. Therefore, the time for performing these experiments is minimized since they only involve selective breeding and not *de novo* generation of transgenic mice.

Based on this approach, very informative experiments have been performed in mice expressing conditionally transforming growth factor (TGF)-β1. Such experiments are worthy of mention because they illustrate the power of this technology in elucidating the precise function of certain genes with an otherwise obscure (often controversial) role in carcinogenesis. TGF-β1 is a secreted growth factor that acts as a tumor suppressor during early stages of carcinogenesis, whereas it appears that it is also capable of promoting tumor progression at later stages. To determine at which stage and by what mechanisms this functional switch occurs, transgenic mice were generated in which TGF-β1 expression could be induced in skin tumors at specific stages. These mice were exposed to chemical carcinogenesis in a manner that allows tumorigenesis to develop in progressive stages from benign papillomas to malignant carcinomas in a controllable manner. It was noted that metastases were induced very rapidly upon TGF-β1 induction in papillomas. This function is in marked contrast to the tumor-suppressive effect when TGF-β1 transgene expression was induced at early stages of the disease. Papillomas developing in the transgenic mice exhibited downregulation of TGF-β receptors and their signal transducer, the *Smads*, and loss of the invasion suppressor E-cadherin/catenin complex in the cell membrane. These molecules were only lost in malignant carcinomas in control mice at a much later stage. Furthermore, these papillomas exhibited increased expression of matrix metalloproteinases and increased angiogenesis. This study suggests that TGF-β1 overexpression may directly induce tumor metastasis by initiating events necessary for invasion. The downregulation of the components of TGF-β signaling in tumor epithelia appears to selectively abolish growth inhibition, thus switching the role of TGF-β1 to a metastasis promoter.

While not formally belonging into this particular category of transgenic constructs, inducible gene expression can also be achieved using endogenous promoters that are activated only at specific developmental stages. A characteristic example is offered by the promoter of the whey acidic protein (WAP), which is active in the secretory epithelium of the mammary gland in the late pregnancy and specifically, during lactation. Driving the expression of oncogenes using this promoter results in the development of lactation-dependent tumors, whereas virgin animals remain tumor-free.

Cre/LoxP System

This approach is based on the property of the viral recombinase Cre, which can excise DNA fragments that are flanked by short nucleotide sequences containing 34 bp and named LoxP. After excision only one LoxP site remains in the DNA. As shown in Fig. 4, the system operates when Cre recombinase and the two LoxP sites flanking a sequence of interest are present in the same animal. This approach can be adapted to generate both knock-in and knock-out animals.

The original description of Cre/LoxP technology as adapted for the generation of mutant animals involved the production of mice carrying the simian virus 40 large tumor antigen (T antigen) gene sequence, separated by a 1.3-kbp Stop sequence that contains elements preventing its expression, flanked by two LoxP sites. These transgenic animals were mated with transgenic mice expressing the Cre recombinase under the control of the murine α-crystallin promoter. All double-transgenic offspring developed lens tumors, consistent with the activation of the α-crystallin promoter in this tissue. Subsequent analysis confirmed that tumor formation resulted from large tumor antigen activation via site-specific, *Cre*-mediated deletion of Stop sequences. While this system is based on transgenesis, it can also be adapted for specific activation or excision of genes, as described below.

"Knock-out" animals can be generated by flanking the gene of interest by two LoxP sites using homologous recombination methodology. Subsequently, the animal is crossed with a transgenic animal which expresses Cre under the regulation of a promoter specifically activated in the tissue(s) in which the gene deletion is aimed. The resulting double-mutant animal will have a specific deletion of the gene of interest in the tissue in which Cre is expressed. In order to generate knock-in animals, the gene of interest is cloned downstream of a translation Stop codon flanked by the LoxP sites. In the presence of Cre, the Stop codon is removed and the gene under investigation is expressed. This approach can be used in order to overexpress specific alleles, functionally simulating the trans-

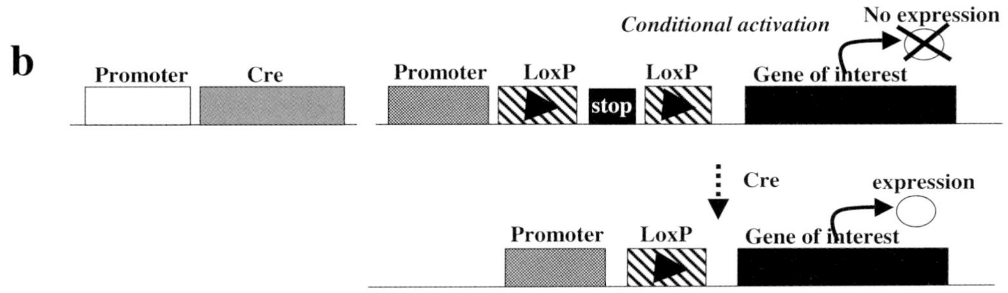

Fig. 4 The Cre/LoxP recombination system for the generation of mutant mice. (a) Using homologous recombination techniques, the gene of interest has been flanked by LoxP sites (floxed) which recombine in the presence of Cre recombinase and are deleted. Cre is expressed by transgenic mice under the regulation of a global or a tissue-specific promoter. (b) For the conditional induction of gene expression a Stop codon is flanked by LoxP sites and the cDNA of interest follows downstream. The Stop codon prevents the translation of the cDNA of interest. In the presence of Cre the Stop codon is deleted and the translation is possible. The whole floxed construct can be introduced either by transgenesis or by homologous recombination.

genic animals in a more precise and controllable manner, or to replace endogenous genes with specific mutant alleles.

In order to achieve overexpression of genes, the homologous recombination is usually targeted towards the ROSA26 locus, which is widely expressed in virtually every tissue in the developing organism through adulthood and its inactivation does not have any physiological consequences.

Inducible Cre/LoxP systems for mice have also been described and they are based on the abovementioned generation of fusion genes with the estrogen receptor. Therefore, following excision by Cre, the gene of interest requires tamoxifen binding in order to be expressed. Generation of transgenic mice expressing a tamoxifen-inducible Cre recombinase protein under the control of the α-myosin heavy chain promoter and breeding with ROSA26 *lacZ-flox*-targeted mice results in the expression of the reporter transgene within the developing and adult heart or the specific, direct recombination and expression of a LoxP-inactivated cardiac transgene in the heart.

Another approach for Cre-mediated activation or inactivation of gene expression is based on the injection, either systemically or locally, of Cre-bearing (expressing) viral constructs into mice that have LoxP sites in the appropriate genetic locus. In this case, Cre-mediated excision does not proceed by a uniform manner as in the cases in which the Cre gene is inherited in the germline, but rather it progresses sporadically in a manner that simulates the development of oncogenic mutations in human tumors, offering the opportunity to develop mouse models of the disease that effectively simulate certain aspects of carcinogenesis in humans.

Note that contrary to tetracycline-based and other analogous systems in which the induction or de-induction of the gene of interest is reversible, the consequences of gene expression are irreversible in the Cre/Lox system animals. Therefore, when the Cre-mediated recombination is accomplished and the induction or deletion of the gene of interest is achieved, it is continued throughout the life of the experimental animal.

Examples of Mouse Models

We now describe some mutant mice models that are of particular interest because they simulate human disease or because they target specific pathways with special value for the elucidation of the basic aspects of carcinogenesis. While this classification is artificial because simulation of the human disease is most frequently the consequence of interference with specific pathways, we find that it is appropriate and helpful for educational purposes.

Pancreatic Tumors

Adenocarcinoma of the pancreas in humans harbors activating mutations in the K-*ras* oncogene in about 80–90% of cases. Furthermore, a considerable amount of these lesions bear inactivating mutations in TSGs, such as p53 and INK4a. Given the high homology between the members of the family, transgenic mice were developed expressing H-*ras* in the exocrine pancreas. These animals developed tumors with acinar characteristics. This study showed that modest amounts of the mutant Ras proteins are sufficient, in an otherwise normal genetic background, to lead to neoplastic transformation of a specific cell type, i.e. differentiating pancreatic acinar cells.

Recently, a more accurate model of the human disease was developed based on the expression of TGF-α under the regulation of the *elastase* promoter. These animals had the *ras* and *Erk1/2* pathways activated through interaction of TGF-α and the epidermal growth factor receptor, and developed premalignant lesions with a tubular structure that resembled the corresponding human lesions. It is notable that when the elastase–TGF-α transgene was introduced into a p53-deficient background, the tumor incidence increased dramatically, therefore simulating the human disease.

Breast Cancer

Mammary cancer models are of particular importance. The mammary epithelium probably represents the tissue for which the largest availability of mutant, cancer-prone animals exists. Two major features of the mammary epithelium are important for this: the availability of the MMTV promoter/enhancer that permits the tissue-specific expression of oncogenes with a high degree of confidence and specificity. Therefore, the development of mice overexpressing mutant alleles that would otherwise be deleterious for the whole organism is feasible for the mammary gland. This organ, even when defective (incapable of lactating), does not considerably affect the survival of the animals, at least in the early stages of the disease, unless breast tumors become metastatic. Furthermore, the mammary gland reaches terminal differentiation during lactation in adult mice, and permits studies aimed at the understanding of the relation between differentiation and carcinogenesis.

Many oncogenes have been expressed in the mammary epithelium and were found capable of inducing the formation of malignant lesions. As an example we have already mentioned *ras* and *myc* oncogenes which induce breast cancers stochastically at a latency that varies between 2 months and 1 year of age. The stochastic manner in which these tumors develop exemplifies the multistage nature of carcinogenesis. An example of a single-step induction of tumors is provided by the expression of the *neu* oncogene under the MMTV promoter in the mammary epithelium. Unlike the stochastic occurrence of mammary cancers in transgenic mice bearing the MMTV/c-*myc* or the MMTV/v-Ha-*ras* transgenes, mice expressing the MMTV/c-*neu* gene uniformly develop mammary adenocarcinomas that involve the entire epithelium in each gland. Considering that these tumors arise synchronously and are polyclonal in origin, expression of the activated c-*neu* oncogene appears to be sufficient to induce malignant transformation in this tissue in a single-step mechanism. In contrast, expression of the c-*neu* transgene in the parotid gland or epididymis, as dictated by the MMTV regulatory elements, leads to benign, bilateral epithelial hypertrophy and hyperplasia which does not progress to full malignant transformation during the observation period. These results suggest that the combination of an activated oncogene and the tissue context are major determinants of malignant progression, and that expression of the activated form of c-*neu* in the mammary epithelium has particularly deleterious consequences.

Interestingly, while all tumors can be classified as adenocarcinomas of the breast, they differ considerably in their pathology, which exhibits an absolute dependency on the initiating oncogenic event. Tumors elicited by activated *ras* consist of small, pink cells with uniform nuclei without significant pleomorphism and relatively abundant cytoplasm, whereas tumors induced by *myc* consist of larger cells and have large pleomorphic nuclei with a coarse, dark chromatin and dark, amphophilic cytoplasm. These tumors have very aggressive, invasive growth patterns. Furthermore, both *ras* and *neu* tumors, contrary to the tumors elicited by *myc* and *wnt*, depend on cyclin D1 expression for their development, as indicated by experiments which involved the introduction of these oncogenes into a cyclin D1-deficient genetic background, and the subsequent evaluation of the latency and incidence of tumor development.

Neurofibromatosis Type I

The product of the NF1 gene encodes for a negative regulator of *ras* and has been found responsible for neurofibromatosis type I – a familial autosomal disease predisposing to the development of neural crest tumors, such as benign neurofibromas and malignant neurosarcomas. Mice with the NF1 gene knocked-out have been developed, and were found to develop tumors such as pheochromocytomas and leukemias, but not neural crest tumors. Although these findings confirm the tumor-suppressive function of NF1, they raised questions regarding the validity of the model. However, neurofibromas did develop when chimeric animals were generated that only partially consisted of NF1-null cells. In addition, mice that carry linked germline mutations in NF1 and p53 develop malignant peripheral nerve sheath tumors (MPNSTs), supporting a cooperative and causal role for p53 mutations in MPNST development. Through use of a conditional (*cre/lox*) allele, it has been demonstrated that loss of NF1 in the Schwann cell lineage is sufficient to generate tumors. In addition, complete NF1-mediated tumorigenicity requires both the loss of NF1 in cells destined to become neoplastic and its heterozygosity in non-neoplastic cells, underlining again the importance of the tumor microenvironment in carcinogenesis. The investigators conclude that the requirement for a permissive haploinsufficient environment to allow tumorigenesis may have therapeutic implications for NF1 and, probably, other familial cancers.

Neurofibromatosis Type II

The NF2 gene encodes for a protein called merlin which is involved in the organization of the actin cytoskeleton, and in the regulation of cell adhesion and migration. NF2 patients inherit a defective p53 allele and after somatic inactivation of the remaining wild-type allele develop tumors of the nervous system. $NF2^{+/-}$ mice develop primarily osteosarcomas, fibrosarcomas as well as hepatocellular carcinomas. These are highly metastatic tumors, contrary to the tumors that develop in other mouse models and which usually exhibit reduced metastatic potential. When a mutant form of merlin was expressed under the regulation of a Schwann cell-specific promoter in transgenic animals the mice developed schwannomas, thus mimicking the clinical characteristics of the disease in humans. The phenotype was similar to that of mice in which the NF2 gene was conditionally inactivated in the Schwann cells.

Tumors of the Intestine

Using random mutagenesis experiments, it was found that mice bearing a nonsense mutation in the adenomatous polyposis coli (APC) gene were generated which develop multiple intestine neoplasia (*Min*). The mutant allele responsible for the disease was termed APC^{min} and its presence resulted in the induction of tumors of the small intestine rather than the colon. Apart from their importance in modeling colon cancer, these mice have increased significance because they offer the opportunity to identify modifier genes for tumor development. For example, *Mom1* was identified as a modifier of *Min* in this model, which affects the multiplicity and size of the tumors that develop in the APC^{min} mice. *Mom1* was mapped in a region of synteny conservation with human chromosome 1p35–36 – a region of frequent somatic loss of heterozygosity (LOH) in a variety of human tumors, including colon tumors. Subsequent studies showed that this particular locus harbors the gene *Pla2g2a* which encodes for secretory phospholipase 2A involved in the biosynthesis of the prostaglandins. It is notable that pharmacological inhibition of prostaglandin biosynthesis also affects the size of the intestinal tumors, consistent with the results of the genetic analyses.

Again, using a combination of genetics and pharmacology to assess the effects of reduced DNA methyltransferase activity on this tumor model, it was shown that a reduction in DNA methyltransferase activity in *Min* mice due to heterozygosity of the DNA methyltransferase gene, in conjunction with a weekly dose of the DNA methyltransferase inhibitor 5-aza-deoxycytidine, considerably reduced the average number of intestinal adenomas in the treated heterozygotes. Therefore, DNA methyltransferase activity substantially contributes to tumor development in this mouse model of intestinal neoplasia. Contrary to the general notion that hypomethylation is associated with carcinogenesis, the results in this experimental system argue against an oncogenic effect of DNA hypomethylation.

Using the murine *villin* promoter, which is active in epithelial cells of the large and small intestine, mice were generated bearing the activated *ras* oncogene [K-*ras*(V12G)] in the intestinal epithelium. More than 80% of these transgenic animals displayed single or multiple intestinal lesions, which ranged from aberrant crypt foci (ACF) to invasive adenocarcinomas. Interestingly, no evidence of inactivating mutations of the TSG *Apc*, the "gatekeeper" in colonic epithelial proliferation, was obtained in this model, notwithstanding the fact that spontaneous mutation of the TSG p53 (a frequent feature in the human disease) was detectable. Thus, it is apparent that this animal model partially recapitulates the genetic alterations found in human colorectal cancer.

Hematopoietic Malignancies

Contrary to most solid tumors that are correlated with an accumulation of genetic alterations in somatic cells, leukemias and lymphomas are characterized by recurrent chromosomal rearrangements, translocations and/or inversions that are rarely identified in human tumors. These genetic alterations mostly result in the overexpression of specific genes or the generation of chimeric, fusion proteins by malignant cells. Such an example is offered by the specific translocation between the BCR and ABL locus, resulting in the generation of the “Philadelphia chromosome”. Two major isoforms of the BCR–ABL chimeric protein have been described, i.e. the p190 and p210 (kDa) forms, due to specific breakpoints of the BCR locus. p210 is the hallmark of the chronic lymphoblastic cases, whereas p190 is detected in about 50% of acute lymphoblastic leukemia (ALL) cases.

Historically, the first attempt to generate a mouse model for leukemia were performed with the generation of transgenic mice expressing the chimeric BCR–ABL(p190) fusion gene that results from the generation of the Philadelphia chromosome. The BCR–ABL cDNA was expressed under the regulation of the metallothionein (MT) promoter. The MT promoter drives expression globally in every tissue in a manner that, while detectable in the absence of exogenous metals, is considerably upregulated when mice are fed with zinc. Consistent with the oncogenic properties of this fusion protein, transgenic animals died of lymphoid or myeloid leukemia 10–59 days after birth. A similar construct producing the BCR–ABL(p210) isoform was also used for the generation of transgenic animals which were found to be inducing the formation of B-ALL after zinc administration.

Another example of a specific translocation associated with a particular type of the disease is offered by the fusion of the retinoic acid receptor (RAR)-α with a variety of genes, most frequently the promyelocytic leukemia (PML) gene, located at chromosome 15q22. Development of MT–PML–RAR-α mice showed that MT-driven PML–RAR-α expression induced liver neoplasms and not leukemias following zinc administration for 5 days.

Taken together, the results collected from transgenic mice bearing RAR-α fusion proteins suggest that the chimeric products, although necessary, are not critical for the development of malignancy. A difference between these transgenic systems and the human conditions that may explain, at least in part, the discrepancies observed is that in the transgenic systems the endogenous gene(s) are expressed, whereas in the human conditions they are reduced to heterozygosity. Indeed, reduction of the endogenous PML gene dosage in PML–RAR-α transgenic mice considerably increased the incidence of the frequency of leukemias and reduced the latency of the onset of the disease. Similar observations were also made with promyelocytic leukemia zinc finger (PLZF)–RAR-α mice.

Although very informative, the conventional transgenic approach described above possesses some limitations. Therefore, additional strategies have been developed. Using the knock-in approach, BCR–ABL(p190) was targeted into the endogenous BCR locus inducing B-ALL, which was found to be very similar to that in humans.

Inducible expression strategies have also been used in the study of leukemias. For example, using the tetracycline-based system, transgenic mice were generated which expressed the c-*myc* oncogene conditionally in a manner whereby tetracycline withdrawal induced transgene expression (*tet-off*). These animals developed malignant T cell lymphomas and AML upon removal of tetracycline. Tissue specificity in this model is conferred by the immunoglobulin heavy chain enhancer and the *Sra* promoter, which dictates the expression of the tetracycline transactivator (rTA). Interestingly, in the aforementioned model and consistent with findings obtained using inducible c-*myc* expression targeted in the mammary epithelium, continuous expression of c-*myc* is mandatory for the maintenance of the malignancy as the tumors regressed upon tetracycline administration and, therefore, c-*myc* de-induction. Similar findings were also obtained using tetracycline-based inducible expression of BCR–ABL transgene that developed B-lymphoid malignancies which regressed upon tetracycline re-administration. As in the case of breast cancer, these experiments demonstrate that despite the accumulation of multiple mutations in malignant cells, targeting specific genetic alterations that are causatively associated with the development of the disease, can produce antitumor effects.

Melanoma

Transgenic and/or mutant mouse models have also been developed for malignant melanomas of the skin. Using the tyrosinase promoter driving the simian virus 40 early-region transforming sequences in the pigment cells of the eye, transgenic mice have been generated that developed ocular and cutaneous melanomas. These tumors were hypomelanotic and were histopathologically similar to the human melanomas. The lesions often originated at a young age, chiefly from the retinal pigment epithelium, although also from the choroid and, rarely, from the ciliary body. The tumors grew aggressively, were highly invasive, and metastasized to local and distant sites. Primary skin melanomas arose later and less frequently than eye melanomas. The authors conclude that for both ocular and cutaneous melanomas, the transgenic mice offer numerous possibilities for experimental study of the mechanisms underlying the formation and spread of melanomas.

Another model for melanoma was based on the identification of mice susceptible to the disease due to an insertion mutation that targeted a gene encoding for *Grm1* (metabotropic glutamate receptor 1) with concomitant deletion of 70 kb of the intronic sequence. The confirmation of the involvement of *Grm1* in melanocytic neoplasia was obtained by the creation of an additional transgenic line with *Grm1* expression driven by the dopachrome tautomerase promoter. The phenotype of the transgenic mice was similar to the original mutant mice, being susceptible to melanoma. In contrast to human melanoma, however, these transgenic mice had a generalized hyperproliferation of melanocytes with limited transformation to fully malignant metastasis.

Targeting Pathways

We have discussed in some detail mouse models of human cancers with an emphasis on the tissue that undergoes malignant transformation. It is worth mentioning, however, some additional cancer models which are of specific interest because of the specific pathway that is altered and the resulting phenotype.

p53-deficient Mice

p53 is the most commonly mutated TSG found inactivated in the majority of primary human tumors. It encodes for a nuclear protein with a fundamental role in regulating apoptosis. p53-deficient mice usually develop lymphomas and die at an early age, after a few months. Furthermore, deletion of p53 dramatically increases the sensitivity of the mice to carcinogens and cooperates with several oncogenes in the induction of tumorigenesis. p53 deficiency was also found to reduce the sensitivity of tumors to anticancer agents. Experiments involving double-mutant mice, bearing inactivating mutations in both p53 and *mdm2*, proved the role of the latter as a negative regulator of the p53 tumor suppressor. *mdm2* null mice are not viable; however, breeding of mice heterozygous for *mdm2* and p53 produces progeny homozygous for both p53 and *mdm2* null alleles. Those animals are viable. Therefore, rescue of *mdm2*$^{-/-}$ lethality in a p53 null background suggests that a critical *in vivo* function of MDM2 is the negative regulation of p53 activity. This experiment illustrates the power of *in vivo* experiments in genetically engineered mice to genetically dissect biochemical pathways in the context of the whole organism.

Genomic Instability

The accumulation of mutations by malignant tumors requires the acquisition of a mutator phenotype which is manifested at two levels – the subchromosomal and the chromosomal level.

Microsatellite Instability

DNA analysis of patients suffering from hereditary nonpolyposis colorectal cancer (HNPCC) revealed that they exhibited an increased incidence of microsatellite mutations in their tumors. This phenomenon was termed microsatellite instability. Subsequent genetic studies showed that the genes responsible for the disease encoded for proteins involved in DNA repair. Thus, patients with defects in these pathways accumulated mutations at a very high incidence, increasing the possibility of the acquisition, and therefore selection, of oncogenic mutations. Subsequent studies showed that virtually every tumor tested exhibited microsatellite instability to some degree.

Animals with these genes, such as MSH2, deleted were subsequently generated, and found to be fertile and developed normally. However, these mice exhibited microsatellite instability and were developing lymphomas at an early age. Other genes involved in DNA repair and frequently mutated in human cancers are MLH1, PMS1 and PMS2. In order to better understand their role in cancer susceptibility, mice deficient for the murine homologs of the human genes MLH1, PMS1 and PMS2 were generated. These mice show distinct tumor susceptibility, most notably to intestinal adenomas and adenocarcinomas, and a different mutational spectra, suggesting that a general increase in replication errors may not be sufficient for intestinal tumor formation and that these genes share overlapping, but not identical, functions.

Chromosomal Instability

Chromosomal instability is manifested by aneuploidy and chromosomal rearrangements. The etiologic association between chromosomal instability and carcinogenesis remains unclear; however, it is believed that it produces gene-dosage effects and interferes with the normal regulation of epigenetic alterations such as DNA methylation.

That defects in these pathways have important developmental consequences probably explains the embryonic lethality of animals with homozygous deletions in genes involved in these processes. As an example we mention the deficiency of BRCA1 and 2 genes, which is lethal. BRCA genes are responsible for a large subset of familial breast cancers in humans, and are believed to play a role in homologous recombination and double-strand break DNA repair. Interestingly, however, heterozygous BRCA-deficient animals do not exhibit increased risk for tumor growth, due to deletion of the wild-type allele as might have been expected. However, when either BRCA1 or 2 are mutated specifically in the mammary epithelium, animals develop breast tumors, which in the case of BRCA1 exhibit a long latency and low frequency. These tumors have allelic deletions in chromosome 11 that harbors p53, implying strong synergistic action between BRCA1 and this TSG. Indeed, conditional mutation of BRCA1 in the mammary epithelium in a p53 heterozygous background accelerated tumorigenesis and resulted in the reduction to homozygosity for the remaining wild-type p53 allele. This example is not the only one that underlines the requirement for p53 deficiency as a necessary alteration to induce carcinogenesis. In a model for medulloblastoma induced by tissue-specific deletion of Rb using Cre expression under the glial fibrillary acidic protein promoter, tumorigenesis was promoted only when p53 was mutated. In this case, although Rb was also deleted in the astrocytes, the cells were unaffected.

Telomeres

As mentioned above, eukaryotic chromosomes have repeated hexanucleotide repeats at the ends which protect them from erosion after several rounds of replication. Telomerase is the enzyme that is responsible for the replication of the telomeric repeats and, while in most adult tissues has been found to be absent, it is reactivated in malignant tissues which require indefinite capacity for replication. Thus, telomerase deficiency predicts resistance in tumorigenesis. Indeed, targeted deficiency to telomerase RNA in mice results in resistance to 7,12-dimethylbenz[*a*]anthracene/12-*O*-tetradecanoyl-phorbol-13-acetate (DMBA/TPA)-induced skin cancers. Furthermore, when these mTERC-deficient animals are introduced in a cancer-prone genetic background, such as that of INK4a/Arf deficiency, it becomes apparent that in later generations the onset of telomere dysfunction is associated with suppression of tumorigenesis and increased survival. However, contrary to this finding that is consistent with the tumor-promoting (permissive) function of telomerase is the observation that mTERC deficiency introduced in *Min* mice accelerates the onset of early-stage adenomas, at least during conditions of mild telomere dysfunction (evidenced by the degree and frequency of chromosomal alterations that are directly associated with telomere erosion).

Issues regarding tissue specificity of tumorigenesis are also raised by this model. p53-deficient animals, despite their increased tumor susceptibility, do not develop epithelial tumors, but lymphomas. In a telomerase activity-deficient background, however, they develop epithelial tumors such as those of the skin, intestine and breast. These findings have important consequences regarding potential therapeutic interventions based on the modulation of telomerase activity.

Mitotic Recombination

An increased incidence of mitotic recombination has been recognized as a characteristic property of malignant cells. While, in addition to the chromosomal loss, it has been considered as a major mechanism by which LOH and reduction to homozygosity is associated with the stimulation of carcinogenesis, little attention has been given to the elucidation of this mechanism (of mitotic recombination). The development of mouse models prone to mitotic recombination, and subsequent tumorigenesis, might provide clues regarding the mechanism underlying these processes. Such a possibility is offered by the development of *Blm* mice which represent a model for Bloom's syndrome. Bloom's syndrome is a disease associated with genomic instability and increased susceptibility to carcinogenesis. Using embryonic stem cell technology, viable Bloom mice have been generated that are prone to a wide variety of cancers. Cell lines from these animals show an increased rate of mitotic recombination. Additional experiments using this model indicate that the increased rate of LOH resulting from mitotic recombination *in vivo* is responsible for the high tumor susceptibility in these mice. Furthermore, specific mutations in the *blm* gene render mice viable, exhibiting increased incidence of mitotic recombination and tumorigenesis. However, when APC-deficient animals were introduced in a *blm*(M3) mutant background, a dramatic increase in tumor incidence was noted, with mitotic recombination being the predominant mechanism by which APC was reduced to homozygosity.

Next Generation of Mouse Models

Despite the fact that the "conventional" approaches mentioned above have proved very informative in understanding basic mechanistic aspects of tumorigenesis, particularly with regard to familial conditions, more precise strategies mimicking human disease must be developed, especially relating to sporadic tumors. Some of these approaches that have only recently appeared in the literature will now be mentioned.

Spontaneous Recombination

A major feature of carcinogenesis is its clonal nature. Consistent with this notion, the tumors develop from a limited number of, most likely a single, mutant cell(s) that acquire(s) proliferative advantage. Then, after several rounds of replication and accumulation of mutations it results in the onset of a malignant tumor. This particular, sporadic nature of the tumors is not properly and accurately addressed by the strategies described above which, with the probable exception of viral-mediated Cre/Lox-based excision, introduce the genetic lesions in a large number of the targeted cells. Jacks and coworkers, in an attempt to overcome this limitation and mimic more precisely the manner in which sporadic tumors develop, introduced an activating mutation in the endogenous K-*ras* locus which could be expressed only after spontaneous homologous recombination. Almost 100% of these animals were found to develop lung carcinoma; in humans, about 40% of these lesions harbor K-*ras* point mutations. Interestingly, these animals do not develop pancreatic or colorectal tumors despite the fact that both lesions bear mutations in the *ras* gene family in humans. This discrepancy in the tissue pattern of the tumors that develop in these animals and human cancers associated with *ras* family gene mutations is most likely attributed to differences in the incidence of somatic recombination in different tissues, which according to this model must occur prior to carcinogenesis and indeed induces it.

Restricted Viral Infections

As mentioned before, the intermediate efficiency of retroviral infections can offer a valuable tool to simulate the clonality of sporadic human tumors. Indeed, transgenic mice can be generated which expresses the avian retrovirus receptor (TVA) in a tissue-specific manner. From those cells, only a limited number can be infected by the TVA-bearing genes of interest, such as specific oncogenes, upon selection of appropriate viral titer. This approach resembles the stochastic manner in which primary tumors develop in humans and has been used for several tumor models. Importantly, we mention an ovarian carcinoma model described recently. The lack of an ovarian-specific promoter rules out the possibility of the development of transgenic/mutant mice using conventional approaches such as those described above. However, when using a less strictly expressed promoter, epression of TVA receptor in primary ovarian cells from p53-deficient animals in culture, followed by paired infections of oncogenes such as *myc*, *Akt* or K-*ras*, it resulted in tumorigenic lesions which resembled the primary ovarian tumors in the stochastic manner they developed. In this model, the ovarian surface epithelium is the precursor tissue for these ovarian carcinomas and the introduction of specific oncogenes causes phenotypic changes in the ovarian surface of epithelial cells. The ovarian tumors induced in mice resembled the primary ovarian carcinomas that develop in humans in terms of their rapid progression as well as intraperitoneal metastatic spread.

Targeting Chromosomal Alterations

Chromosomal rearrangements represent a major feature of carcinogenesis; however, most approaches only target specific loci. Taking advantage of the Cre/Lox technology and the fact that Cre can induce the recombination of LoxP sites separated by about 60 cM, animals in which Cre-mediated excision can induce the formation of balanced translocations started to be developed. By performing consecutive targeting of LoxP recombination substrates to the end points of a genetic interval followed by Cre-induced recombination, defined deficiencies, inversions and duplications extending to 3–4 cM can be constructed in embryonic stem cells. Duplication and deletion alleles can subsequently be transmitted into the mouse germline. These strategies are believed to offer great opportunities to design chromosomal alterations with precise junctions that can be very informative in studying their effects in carcinogenesis.

High-throughput Engineering of the Mouse Genome

While very informative, targeting specific mutations in the mouse genome is both time consuming and quite laborious. However, recent advances in genetic manipulation of mice permit the development of high-throughput and largely automated processes that can accelerate the mutagenic process. Such an approach has recently been developed, termed "VelociGene", and uses targeting vectors based on bacterial artificial chromosomes (BACs). This approach permits genetic alteration with nucleotide precision and can precisely replace the gene of interest with a reporter that allows for high-resolution localization of target gene expression.

Bibliography

Akagi K, Sandig V, Vooijs M, Van der Valk M, Giovannini M, Strauss M, Berns A. Cre-mediated somatic site-specific recombination in mice. *Nucleic Acids Res* **1997**, *25*, 1766–1773.

Bernardi R, Grisendi S, Pandolfi PP. Modelling haematopoietic malignancies in the mouse and therapeutical implications. *Oncogene* **2002**, *21*, 3445–3458.

Burns PA, Bremner R, Balmain A. Genetic changes during mouse skin tumorigenesis *Environ Health Perspect* **1991**, *93*, 41–44.

Callahan R. MMTV-induced mutations in mouse mammary tumors: their potential relevance to human breast cancer. *Breast Cancer Res Treat* **1996**, *39*, 33–44.

Chang S, Khoo C, DePinho RA. Modeling chromosomal instability and epithelial carcinogenesis in the telomerase-deficient mouse. *Semin Cancer Biol* **2001**, *11*, 227–239.

Cichowski K, Shih TS, Schmitt E, Santiago S, Reilly K, McLaughlin ME, Bronson RT, Jacks T. Mouse models of tumor development in neurofibromatosis type 1. *Science* **1999**, *286*, 2172–2176.

Dooldeniya MD, Warrens AN. Xenotransplantation: where are we today? *J R Soc Med* **2003**, *96*, 111–117.

Fisher GH, Orsulic S, Holland E, Hively WP, Li Y, Lewis BC, Williams BO, Varmus HE. Development of a flexible and specific gene delivery system for production of murine tumor models. *Oncogene* **1999**, *18*, 5253–5260.

Fukumura D, Xavier R, Sugiura T, Chen Y, Park EC, Lu N, Selig M, Nielsen G, Taksir T, Jain RK, Seed B. Tumor induction of VEGF promoter activity in stromal cells. *Cell* **1998**, *94*, 715–725.

Geng Y, Whoriskey W, Park MY, Bronson RT, Medema RH, Li T, Weinberg RA, Sicinski P. Rescue of cyclin D1 deficiency by knockin cyclin E. *Cell* **1999**, *97*, 767–777.

Gossen M, Bujard, H. Tight control of gene expression in mammalian cells by tetracycline-responsive promoters. *Proc Natl Acad Sci USA* **1992**, *89*, 5547–5551.

Jacks T, Weinberg RA. Taking the study of cancer cell survival to a new dimension. *Cell* **2002**, *111*, 923–925.

Jackson-Grusby L. Modeling cancer in mice. *Oncogene* **2002**, *21*, 5504–5514.

Lewandoski M. Conditional control of gene expression in the mouse. *Nat Rev Genet* **2001**, *2*, 743–755.

Muller WJ, Sinn E, Pattengale PK, Wallace R, Leder P. Single-step induction of mammary adenocarcinoma in transgenic mice bearing the activated c-*neu* oncogene. *Cell* **1988**, *54*, 105–115.

Sinn E, Muller W, Pattengale P, Tepler I, Wallace R, Leder P. Coexpression of MMTV/v-Ha-*ras* and MMTV/c-*myc* genes in transgenic mice: synergistic action of oncogenes *in vivo*. *Cell* **1987**, *49*, 465–475.

Valenzuela DM, Murphy AJ, Frendewey D, Gale NW, Economides AN, Auerbach W, Poueymirou WT, Adams NC, Rojas J, Yasenchak J, Chernomorsky R, Boucher M, Elsasser AL, Esau L, Zheng J, Griffiths JA, Wang X, Su H, Xue Y, Dominguez MG, Noguera I, Torres R, Macdonald LE, Stewart AF, DeChiara TM, Yancopoulos GD. High-throughput engineering of the mouse genome coupled with high-resolution expression analysis. *Nat Biotechnol* **2003**, *21*, 652–659.

Van Dyke T, Jacks T. Cancer modeling in the modern area: progress and challenges. *Cell* **2002**, *108*, 135–144.

19
Multistage Carcinogenesis in Humans: Molecular Epidemiology and the Colon Cancer Model

Research data derived from *in vitro* experiments in tissue culture and *in vivo* experiments in mice and other organisms would not have any meaning in the cancer field if researchers were not able to demonstrate their biological relevance. Therefore, it is imperative to show that these alterations indeed occur in human lesions and that they contribute to the development of the disease. A human cancer that has been extensively studied at the molecular level is colon cancer. Various genetic lesions have been identified in this type of human malignancy and were shown to occur not randomly, but rather deterministically in sequential order, in a manner that is considered necessary for the progression of the disease.

Almost two decades ago Vogelstein and coworkers proposed a model for colon cancer development after analyzing in great detail, at least at the level permitted by the molecular technology of the time, the tumor DNA from colon cancer lesions. In this model the progressive histopathology of the corresponding lesions was associated with specific changes in oncogenes and tumor-suppressor genes (TSGs) that were considered as prerequisites for the transition of the disease at the next stage. While the detection and subsequent evaluation of the incidence of specific alterations at the DNA of various different tumors had been successfully performed at that time, the importance of the their model is due to the fact that a unified hypothesis was proposed that dynamically associates the specific molecular lesions with the phenotype of the disease. Furthermore, the sequential order in which these lesions should have occurred was demonstrated, linking genetics with histopathology and probably introducing the field of molecular pathology.

Since that time Vogelstein's model for colon cancer progression, also called the "Vogelgram" by some, serves as a paradigm for multistage carcinogenesis in other tumors as well; while has been enriched and extended by additional information, its major concept remains intact.

The Model

The model proposed by Fearon and Vogelstein back in 1990 is summarized in Fig. 1. Four major players have been recognized to play a causative role in colorectal cancer development according to this model, and it involves inactivation of

Understanding Carcinogenesis. Hippokratis Kiaris

ISBN 3-527-31486-5

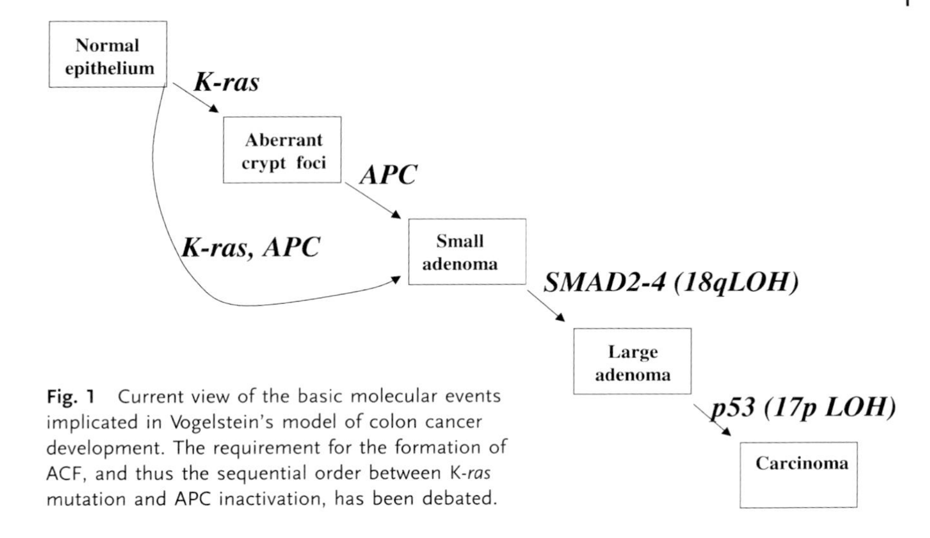

Fig. 1 Current view of the basic molecular events implicated in Vogelstein's model of colon cancer development. The requirement for the formation of ACF, and thus the sequential order between K-*ras* mutation and APC inactivation, has been debated.

the adenomatous polyposis coli (APC) tumor suppressor, activation of the K-*ras* oncogene, and deletions at chromosomal arms 18q and 17p.

The earliest hit in colorectal cancer formation is thought to be the mutational inactivation of the APC TSG, which results in the formation of (early) adenomas. Most likely APC inactivation deregulates the stem cell compartment of the colon and is involved in *wnt* signaling. In normal cells, with intact APC function, APC regulates β-catenin levels. Upon its inactivation, β-catenin accumulates in the cytoplasm and translocates into the nucleus where it activates the transcription of downstream targets genes, such as the potent oncogene c-*myc* and the cycle regulator cyclin D1, by a mechanism that may involve the TCF-4 transcription factor.

It is noteworthy that germline mutations in the APC gene are responsible for the syndrome familial adenomatous polyposis (FAP), which is associated with the dominant predisposition to hundreds of adenomatous polyps in the colon and rectum as well as upper gastrointestinal tract tumors. Thus, the same lesion that is thought to be responsible for the development of sporadic colorectal cancer, when inherited in the germline, also results in the development of similar lesions with a comparable histopathology and molecular profile. The development of mice with germline mutations in the APC gene confirmed the observations and provided a model to identify modifiers for colon cancer development (see Chapter 18).

A subsequent event (today we think that it may also correspond to a primary initiating event in some cases) is the mutational activation of the oncogene K-*ras*. K-*ras* mutations are thought to accompany the transition into a large adenoma. This is achieved by the ability of mutant K-*ras* to further increase proliferation and to inhibit apoptosis by a mechanism that involves the stimulation of the activity of mitogen-activated protein kinase, Raf, phosphatidylinositol-3-kinase/

AKT, etc., pathways. The molecular events that follow and are considered responsible for the induction of the malignant conversion are chromosomal deletions at 18q and 17p. Regarding the specific targets of these alterations, 17p deletion involves the p53 tumor suppressor that among its various consequences results in the abolishment of DNA damage-induced checkpoints, thus inhibiting apoptotic cell death and increasing genomic instability. The resulting malignant lesions that have increased genomic instability can also be metastatic. With regard to deletions at 18q, they were originally thought to target a TSG designated DCC (deleted in colorectal cancer), although subsequently it turned out that they target a locus encoding for netrin-1 that is involved in cell adhesion. However, subsequent studies implicated additional genes in the 18q arm as the major targets of the 18q deletions, including SMAD-4 that is involved in transforming growth factor (TGF)-β signaling.

Collectively, these alterations induce the accumulation of additional changes in the genome/expression profile of the targeted cells that result in the development of the fully fledged malignancy and the acquisition of the characteristics that constitute the hallmarks of cancer. These alterations have been described by Hahn and Weinberg (2000), and include autonomy in growth signals, resistance to antigrowth signals, limitless potential for cell division, evasion of apoptosis, angiogenesis, and the ability to invade and metastasize.

Colon Cancer Model Revisited

Strict interpretation of the aforementioned model predicts that the corresponding genetic lesions, i.e. inactivation of the APC tumor suppressor, the activation of K-*ras* oncogene, and deletions at chromosomal arms 18q and 17p, should be present concomitantly in malignant colorectal cancers in the vast majority, if not all, of the cases. However, it has been found that this true for less than 10% of the cases, suggesting the operation of alternative pathways that drive colorectal cancer development. Indeed, mutations in the oncogene BRAF and the TGF-β receptor II can substitute for the "typical" mutations in K-*ras* and Smads, especially in tumors characterized by microsatellite instability due to defects in DNA repair. Furthermore, mutations in the K-*ras* oncogene have been detected in both normal colonic mucosa and in aberrant crypt foci (ACF). Among the latter, almost 80–90% of nondysplastic ACF bear K-*ras* mutations – a rate which is reduced to about 57% for the dysplastic ACF. This observation is in contrast to the notion that dysplastic ACF represent the precursors of adenomas, and again implies the operation of alternative pathways and reduced impact of the K-*ras* mutations in the development of colorectal cancer. It has to be emphasized that in a given tumor, when we evaluate the impact of certain lesions and their contribution to the development of the disease, identification of genetic alterations other than those described above does not necessarily mean the operation of alternative pathways, but may also reflect the occurrence of genetic hits that despite their stabilization in the genome are not associated with carcinogenesis in a causative manner.

Genomic Instability in Colorectal Cancer Development

It has been mentioned in the description of the genes affecting the fidelity of DNA replication that the induction at some degree of genomic instability, termed the acquisition of a "mutator phenotype", is necessary for the progressive accumulation of mutations during carcinogenesis. This phenomenon has been studied quite extensively in cancers of the colon and led to the classification of these cancers into two major categories – MIN (microsatellite instability) cancers covering about 15% of the cases and CIN (chromosomal instability) cancers covering approximately 85% of the cases. The first type of cancers (MIN classification) is due to mutations in DNA repair genes (MSH2, MSH6, MLH1 and others) that result in replication errors that are reflected in the formation of mutations in microsatellite DNA sequences. Tumors that belong to the CIN category bear gross chromosomal alterations. The mechanism of the induction of chromosomal instability remains less understood, but is thought to involve mitotic checkpoint genes such as BUB1. The "Vogelstein model" discussed previously appears more applicable to the CIN tumors. In addition, epigenetic changes in the DNA, although quite common in colorectal carcinogenesis, are not discussed in this model, but it is proven that they play an important role during the progression of the disease. One example is offered by the hypermethylation of DNA repair genes, such as MLH1, that results in their silencing and eventually in the induction of genomic instability.

Other Tumor Models

Following the development of the colorectal cancer model, investigators have performed molecular analyses in other types of cancers that led to the suggestion of different models for the carcinogenesis in a variety of cancers. For example, cancer of the prostate represents a relatively well-studied cancer at the molecular level (Fig. 2). 8p deletions have been recognized as an early event that occurs in prostatic intraepithelial neoplasia (PIN) and in prostatic inflammatory atrophy (PIA). This genetic lesion, which persists at more advanced stages of the disease since it can be detected in the majority of metastatic prostatic cancers, is thought to target the homeobox gene NKX3.1 (or NKX3A). This gene is induced by androgens and, in turn, stimulates the expression of various prostate-specific genes. The importance of NKX3.1 is exemplified by the fact that its deletion in mice results in inhibition of prostate maturation and hyperproliferation. However, a direct tumor-suppressive function remains to be established.

An early genetic lesion for prostate cancer development is also the upregulation of the recently discovered marker α-methylacyl-CoA racemase (AMACR) that catalyzes the conversion of *R*- to *S*-stereoisomers of branched chain fatty acids, facilitating their metabolism via β-oxidation. It is notable that genetic variants representing either innocent polymorphisms or actual mutations of this gene have also been identified in patients suffering from some hereditary forms of prostate can-

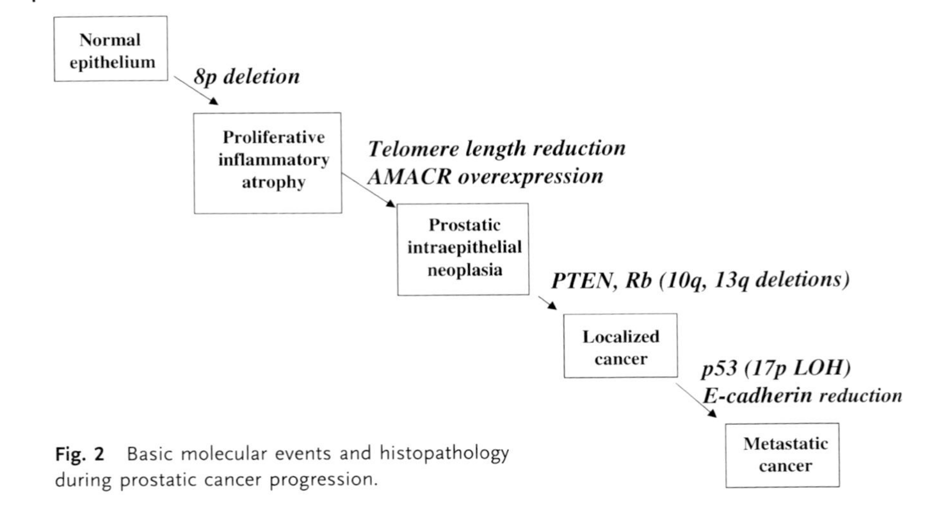

Fig. 2 Basic molecular events and histopathology during prostatic cancer progression.

cer. Decrease in telomere length has also been reported in PIN lesions and is thought to precede the upregulation of telomerase activity that occurs at more advanced stages.

At subsequent stages, allelic deletions at 10q that target PTEN (phosphatase and tensin homolog deleted on chromosome 10) and 13q that target the Rb TSGs occur, and have been associated with the progression of the disease. However, whether Rb deletion is decisive for the progression of prostate cancer has not yet been established. Additional alterations that can be detected at these stages and play a role in the acquisition of a more aggressive phenotype involve the suppression of the expression of the cell cycle regulator p27. Finally, when the tumor becomes metastatic, deletions at 17p affecting the p53 tumor suppressor have been observed as well as downregulation of E-cadherin and increased DNA methylation, which results in the silencing of TSGs. An important lesion for the biology of prostatic carcinomas involves the acquisition of a hormone (androgen)-independent phenotype that is due to the loss of the androgen receptors from the cancer cells. This latter alteration underlines the aggressive behavior of prostatic carcinomas and has important implications for the management of the disease. In a relevant model of the disease this transition of the prostate cancer from the androgen-dependent into the androgen-independent state involves the change of the relative quantity of prostate cancer stem cells which are androgen-independent cells versus differentiated cancer cells that remain androgen-dependent cells. Upon androgen depletion, differentiated cells are reduced and are substituted by the stem cells in which, since they need to proliferate, the probability of acquiring mutations that confer androgen independence increases.

Relatively recently, a model for ovarian carcinoma has also appeared in the literature. According to this model proposed by Shih and Kurman (2004), endometrial carcinomas can be divided into two major categories – type I and II tumors.

Type I tumors arise from borderline tumors and are generally low-grade neoplasms, whereas type II tumors are high-grade neoplasms with unknown precursor lesions. Molecular alterations that define type I tumors are mutations in K-*ras* and BRAF oncogenes that importantly are mutually exclusive, suggesting that they have similar effects in carcinogenesis. These alterations occur early in the disease because they can be detected in small atypical proliferative serous tumors. In contrast to type I tumors that only display a limited incidence of p53 mutations, type II tumors are defined by mutations in the p53 tumor suppressor. Genomic instability, manifested by the degree of allelic imbalance, is detectable in both types of tumors and increases progressively in low-grade tumors, but is high in high-grade tumors even at early stages.

Implications of "Modeling" Cancer

Understanding the sequential steps that are required in carcinogenesis in a given tissue will make intervention feasible by developing drugs that target pathways that are specifically activated (or inactivated) at particular stages. This will result in slowing down the rate of disease development and prolongation of the patient's life. Furthermore, assessment of the specific molecular changes in the patient's tumor DNA may permit a more accurate prediction for the prognosis of the disease that will be based on the specific genetic lesions carried by the tumor rather than its histopathology, which is conceivably the result of these alterations.

Bibliography

Arends JW. Molecular interactions in the Vogelstein model of colorectal carcinoma. *J Pathol* **2000**, *190*, 412–416.

Bellacosa A. Genetic hits and mutation rate in colorectal tumorigenesis: versatility of Knudson's theory and implications for cancer prevention. *Genes Chromosome Cancer* **2003**, *38*, 382–388.

Fearon ER, Vogelstein B. A genetic model for colorectal tumourigenesis. *Cell* **1990**, *61*, 759–767.

Fodde R, Smits R. Disease model: familial adenomatous polyposis. *Trends Mol Med* **2001**, *7*, 369–373.

Gonzalgo ML, Isaacs WB. Molecular pathways to prostate cancer. *J Urol* **2003**, *170*, 2444–2452.

Schulz WA, Burchardt M, Cronauer MV. Molecular biology of prostate cancer. *Mol Hum Reprod* **2003**, *9*, 437–448.

Breivik J, Gaudernack G. Genomic instability, DNA methylation, and natural selection in colorectal carcinogenesis. *Semin Cancer Biol* **1999**, *9*, 245–254.

Shih IM, Kurman RJ. Ovarian tumorigenesis. A proposed model based on morphological and molecular genetic analysis. *Am J Pathol* **2004**, *164*, 1511–1518.

Vogelstein B, Fearon ER, Hamilton SR, Kern SE, Preisinger AC, Leppert M, Nakamura Y, White R, Smits AM, Bos JL. Genetic alterations during colorectal-tumor development. *N Engl J Med* **1988**, *319*, 525–532.

Part V
Future Perspectives

Understanding Carcinogenesis. Hippokratis Kiaris

ISBN 3-527-31486-5

20
Epilogue

The applications of the recent advances of tumor biology in oncology have already been mentioned throughout this book. Here, in summary, we will mention again the three major directions of cancer research, i.e. diagnosis, prognosis and therapy. The importance of these advances is due to the fact that we are now able to look and assess the causes at the molecular level, rather than the symptoms of the disease.

Diagnostically, detection of alterations at the DNA level or the expression level of specific markers permits the identification of cancer cells much earlier than at the stage where the disease is clinically detectable (symptomatic). The efficiency of therapy increases dramatically at such an early stage. Virtually all types of DNA alterations constitute candidates that can be used to detect cancer at an early stage. Each of them, e.g. microsatellite alterations and oncogene/TSG mutations, have their own advantages and disadvantages. Considering the high heterogeneity of cancer, both at the level of the specific pathways engaged in its progression and its clinical manifestation, a combination of such tests should be used to accurately identify precancerous lesions.

Prognosis of the outcome of the disease, including that of the efficiency of specific anticancer therapies, also represents a target that is feasible to achieve in the near future. Again, the advantage of such an approach is that its determination, contrary to conventional approaches based on the correlative assessment of specific clinicopathological parameters, now relies on the evaluation of the status of specific genetic factors that directly dictate the tumor's behavior. For example, instead of evaluating the aggressive histopathology that may partially be due to p53 mutations, as a marker for reduced sensitivity against certain therapies, we may able to directly predict the therapeutic response in association with the p53 status.

In the whole field of cancer, *therapy* targeting the causes of cancer, at the molecular level, by correcting the altered function of cancer-associated gene(s) or by using molecular advances to achieve efficient tumor targeting, probably represents the most challenging goal of cancer biology. By definition, many of the genetically altered mouse models in which modification of the genetic background changes the profile of carcinogenesis provide the rational for the development of gene-based therapies.

Understanding Carcinogenesis. Hippokratis Kiaris

ISBN 3-527-31486-5

Finally, in addition to this directly applicable dimension of cancer research, a more long-term contribution from studying tumor biology emerges. While cancer research is, or should be, by definition applicable, it should always be kept in mind that a tremendous amount of our knowledge regarding cellular physiology and molecular biology has been derived by cancer researchers studying the causes of malignancy. Thus, this particular contribution of cancer science should not be ignored.

Index

Understanding Carcinogenesis. Hippokratis Kiaris

ISBN 3-527-31486-5

c

d

e